T0243059

CAMBRIDGE LIBRARY COLLECTION

Books of enduring scholarly value

Botany and Horticulture

Until the nineteenth century, the investigation of natural phenomena, plants and animals was considered either the preserve of elite scholars or a pastime for the leisured upper classes. As increasing academic rigour and systematisation was brought to the study of 'natural history', its sub-disciplines were adopted into university curricula, and learned societies (such as the Royal Horticultural Society, founded in 1804) were established to support research in these areas. A related development was strong enthusiasm for exotic garden plants, which resulted in plant collecting expeditions to every corner of the globe, sometimes with tragic consequences. This series includes accounts of some of those expeditions, detailed reference works on the flora of different regions, and practical advice for amateur and professional gardeners.

A Treatise upon Planting, Gardening, and the Management of the Hot-House

This practical guide was first published in 1776: in this reissue of the two-volume 1777 second edition, the two volumes have been bound in one book. John Kennedy (d.1790) was the gardener to Sir Thomas Gascoigne, the owner of Parlington Hall in Yorkshire, and his book is addressed to landowners and their head gardeners. His concern is with 'the planting of poor wastes, moorlands, and apparent mountains', as well as with hothouse plants such as pineapples and vines, and delicacies including asparagus and cultivated mushrooms. At the other end of the scale, he also provides sections on field-cabbages, carrots and turnips as feed for cattle. In each of these areas, he gives detailed descriptions of the preparation of the ground, the tools needed, propagation techniques, and the subsequent management of pests and diseases. This is a fascinating treatise on the gardening skills needed on a grand eighteenth-century agricultural estate.

Cambridge University Press has long been a pioneer in the reissuing of out-of-print titles from its own backlist, producing digital reprints of books that are still sought after by scholars and students but could not be reprinted economically using traditional technology. The Cambridge Library Collection extends this activity to a wider range of books which are still of importance to researchers and professionals, either for the source material they contain, or as landmarks in the history of their academic discipline.

Drawing from the world-renowned collections in the Cambridge University Library and other partner libraries, and guided by the advice of experts in each subject area, Cambridge University Press is using state-of-the-art scanning machines in its own Printing House to capture the content of each book selected for inclusion. The files are processed to give a consistently clear, crisp image, and the books finished to the high quality standard for which the Press is recognised around the world. The latest print-on-demand technology ensures that the books will remain available indefinitely, and that orders for single or multiple copies can quickly be supplied.

The Cambridge Library Collection brings back to life books of enduring scholarly value (including out-of-copyright works originally issued by other publishers) across a wide range of disciplines in the humanities and social sciences and in science and technology.

A Treatise upon Planting, Gardening, and the Management of the Hot-House

JOHN KENNEDY

CAMBRIDGE
UNIVERSITY PRESS

CAMBRIDGE
UNIVERSITY PRESS

University Printing House, Cambridge, CB2 8BS, United Kingdom

Cambridge University Press is part of the University of Cambridge.

It furthers the University's mission by disseminating knowledge in the pursuit of
education, learning and research at the highest international levels of excellence.

www.cambridge.org
Information on this title: www.cambridge.org/9781108072243

This edition first published 1777
This digitally printed version 2014

ISBN 978-1-108-07224-3 Paperback

A
TREATISE
UPON
PLANTING,
GARDENING,
AND THE
MANAGEMENT OF THE HOT-HOUSE.

CONTAINING

I. The Method of planting Foreſt-Trees in gravelly, poor, mountainous, and heath Lands; and for raiſing the Plants in the Seed-Bed, previous to their being planted.

II. The Method of Pruning Foreſt-Trees, and how to improve Plantations that have been neglected.

III. On the Soils moſt proper for the different Kinds of Foreſt-trees.

IV. The Management of Vines; their Cultivation upon Fire-Walls and in the Hot-Houſe; with a new Method of dreſſing, planting, and preparing the Ground.

V. A new and eaſy Method to propagate Pine Plants, ſo as to gain Half a Year

in their Growth; with a ſure Method of deſtroying the Inſect ſo deſtructive to Pines.

VI. The beſt Method to raiſe Muſhrooms without Spawn, by which the Table may be plentifully ſupplied every Day in the Year.

VII. An improved Method of cultivating Aſparagus.

VIII. The beſt Method to cultivate Field Cabbages, Carrots, and Turnips for feeding of Cattle.

IX. A new Method of managing all Kinds of Fruit-Trees, viz. of proper Soils for planting, of pruning and dreſſing them; with a Receipt to prevent Blights, and cure them when blighted.

By JOHN KENNEDY,
GARDENER TO SIR THOMAS GASCOIGNE, BART.

THE SECOND EDITION,
CORRECTED AND GREATLY ENLARGED.

IN TWO VOLUMES.

VOL. I.

LONDON:
PRINTED FOR S. HOOPER, Nº 25, LUDGATE-HILL; and ſold by G. ROBINSON, PATERNOSTER-ROW.

M DCC LXXVII.

T O

Sir THOMAS GASCOIGNE, Bart.

THE fruits of my labours, whatever they may be, belong to you: your goodnefs and generofity have already greatly over-paid them; yet I never durft have petitioned for your patronage and protection for thefe Sheets, which I prefume to offer to the public, but from a perfect conviction of their utility: being fufficiently acquainted with your knowledge in my profeffion, and more fo with the rigour with which you would treat the errors of

your

iv DEDICATION.

your own servant when they regarded the public.

I have the honour to be, with the
utmost respect and gratitude,

SIR,

Your dutiful servant,

JOHN KENNEDY.

PREFACE.

A Second Publication of this Work affords the Author the pleasing task of returning thanks to the Public for the favorable reception it has met with, which gives him great pleasure and satisfaction.

He thinks himself obliged, in a particular manner, to those Gentlemen who have honored him with letters of approbation of his Work, many of whom have generously confessed the singular advantage their Pineries have received by applying his preparation for destroying those insects so pernicious and destructive to the Ananas or Pine-Apple, which it has effectually completed.

Many purchasers of the first edition of this work, as well as his friends,

a 3   having

having requested him to publish his
thoughts upon the Management and
Pruning of Fruit-trees in general,
assuring him that it would prove a
valuable addition to this work.

In compliance with their flattering
request he has ventured to lay before
the Public a course of many years
practice, which he has followed with
the greatest success.

The many publications on Gardening and Planting, which have been
offered of late years to the Public,
might have discouraged the Author
of this Treatise from the present
attempt; but as most of those that
have fallen in his way treat the subject
in too general and speculative a method
to be of service to practitioners, his
intention in the following sheets is
not to deliver himself systematically,
but, in the most explicit manner, to

5   lay

lay before the Public facts that have been fuccefsfully reduced to practice by himfelf.

Each particular fubject he means to treat of, he will take up from the beginning, and to give the moft minute directions as to the method of culture, labour, and management; together with the feafons that each particular work is to be performed in.

The planting of poor waftes, moorlands, and apparent barren mountains, has been but feldom treated of, and in very few places attempted.

The fuccefs the Author has had in planting fuch grounds, even in the north of Scotland, has induced him to treat that fubject rather largely; and he flatters himfelf that, if his directions are followed, extenfive tracts of land which are now ufelefs, may become ornamental and profitable,

viii   PREFACE.

A general fyftem of Gardening not being the intention of this Treatife, the Author will confine himfelf to the management of Fruit-trees in general, of Vines, Ananas or Pine-Apple, Afparagus, and a new method of raifing Mufhrooms without fpawn.

His method of managing the different Soils proper for Borders, for Planting Fruit-trees, and of Pruning them, if not the beft, has at leaft the merit of being fo new, that it differs entirely from any thing he has feen, both as to theory and practice.

The management of Fruit-trees has been treated at large by moft of the eminent writers on the fubject of Gardening. The Author has read them with attention, and while he fees great objections to all their plans, finds lefs reafon to attach himfelf to any one of them, as there are hardly

two

two writers of one opinion; but it is his bufinefs to endeavour to be as accurate as poffible himfelf, and not to criticize on the defects of others.

He would not however have it underftood, that by reading this treatife every perfon who can diftinguifh a peach from a pear, will be able to undertake the management of Fruit-trees, &c. for if books could make proficients, there would be no occafion for mafters in any art or branch whatever.

Neither fhould any one attempt to plant or prune without having been fome time under a fkilful perfon, whofe practical inftruction being added to a careful obfervation of the rules here given, cannot fail of fuceeeding to the utmoft expectation of thofe who follow them, which is the moft ardent wifh of the Author.

The

The directions given on those heads being very different from the general practice, may perhaps make some rather diffident in following them; but the Author avers that they are what he has followed with the greatest success for many years.

Agriculture being now the object of general attention, the Author has added to this Treatise the cultivation of Field-Cabbages and Carrots, induced thereto by the great crops he has himself raised, and the great advantage they are in feeding of cattle, &c.

The growing of Turnips too is become an object of the greatest consequence for feeding of Cattle; the Author, therefore, has from many accurate observations added a chapter on their cultivation; his method has been followed by many who have raised them
with

with the greateſt ſuccefs for years in different parts of England. The great waſte every year, occaſioned by their rotting (and the larger they are the more they are liable to it) muſt be a conſiderable loſs to the grower, which may be avoided in a great meaſure by pulling and houſing them, eſpecially the large ones, in dry weather, which, from repeated trials he is warranted to ſay, will effectually preſerve them.

The inſtructions given in this Treatiſe upon Planting, Gardening, and rural Oeconomy, are the reſult of many years experience; and the approbation his labours have met with is a very ample reward to the Author.

CONTENTS

OF

VOLUME THE FIRST.

A

TREATISE

ON

PLANTING

AND

GARDENING.

CHAP. I.

The Method of raising wood on rocky, hilly, waste, and heath lands.

THE extensive tracts of rocky, waste, and heath lands in this kingdom, if converted into plantations of thriving trees, would prove a certain benefit to posterity, as well as a pleasing reflection to those who are at the expence of performing so great and good a work. To assist the generous planter in

his

his patriotic defign, is the intention of this firſt chapter. I ſhall ſuppoſe the tract of country, now in view, to conſiſt of rocky, hilly, and heath land with little or no ſoil. Such unpromiſing ground may be covered with wood in the following manner.

THE labourers ſhould be provided with light ax-hoes, broad at one end and narrow at the other. With the broad end pare off the graſs or heath as thin as poſſible ; then with the narrow end ſtir the ground to four or five inches, if you can; picking out ſuch ſmall ſtones as are looſened by the hack in ſtirring the ground, always avoiding large ſtones ; but where theſe, or pieces of rock intervene, at three feet diſtance from the rock make as many places round as you can, for no ground ſhould be loſt; and although the trees are near on one ſide, they will have ſufficient air to grow to maturity.

THIS work ſhould be done in ſummer,
that

that the places loofened may have the advantage of the firft rains in autumn to moiften them; for there is no planting in fuch grounds until they are moiftened after ftirring, as all rain runs off before the furface is broke.

THERE is nothing more to be done until the planting-feafon, which fhould be as early as the weather will permit; for if done late, a dry fpring would be of bad confequence.

WHEN you begin to plant, take up no more trees at a time than can be planted in one day, taking care not to expofe the roots to the fun or wind.

THE beft method is to fow and plant the trees alternately. So when you plant, the places for feed fhould be left until the feafon for fowing in fpring.

THE trees fit for planting in fuch places are, at the fummit of hills, Scotch firs and
A 2   larches.

larches. This is the proper fituation for larch, it being an inhabitant of high and cold places. A great reafon for its growing crooked, is its being planted in low fituations and good land, where it grows too faft, and is not able to fupport its large head.

The weft, north, and north-eaft afpects fhould be planted with Scotch fir and larch; and towards the bottom of the hill, in the fame afpects, beech will thrive. If there are fix inches of foil, fown or planted oaks will grow very well; and though the foil be poor, clumps of fyca-more for ornament will grow beyond ex-pectation; as they will receive the moifture from the higher grounds. This may be feen in natural woods.

The other afpects fhould be planted with beech, hornbeam, fycamore, and all the bottom with oaks: if three or four mountain afhes are planted in dif-ferent places, it will add to the beauty of the

the plantation; but the wood is of little value.

THE common wood, or rough-leaved Elm, will grow in a very poor foil to great perfection, and may be planted next to the Beech. Thefe muft be planted very thick. There are many firs and pines brought from America that thrive in poor land. Of thefe there have been no large plantations made; what have been planted are for ornament, and the wood of them does not feem to differ much from the Scotch, which we are fure comes to a great fize in a very poor foil, and at a diftance they have much the fame appearance. The beauty of fuch plantations is only to be feen from diftant views. The feeds of the fame fort of trees fhould be fown in every other place left vacant in planting. At the top, where the Scotch firs and larches are planted, there fhould be no places left, as thefe trees are of a very quick growth, and the feeds of the fir kinds are fubject to be

A 3   de-

devoured by birds. The young plants alfo, for the firft year, are very fubject to be thrown out of the ground by froft, And, what is more material, they would be fmothered by thofe planted, if they fhould meet with no mifchance.

In fowing, you may put fome acorns amongft the planted beech, as they are near of a growth. They will grow from feed where they will not thrive when planted, and, penetrating into the cavities with their young fibrous roots, will find fufficient nourifhment where there is little appearance of a tree's growing. In natural woods we often fee fine oaks in fuch fituations, and there is no doubt fuch trees were from feeds accidentally dropt.

It fhould be obferved, that all trees thrive better in clumps than when mixed. If mixed, they fhould be with trees of an equal growth, which is feldom confidered. It has been a common practice to mix Scotch firs with oak and beech (the Scotch

Scotch fir is of a very quick growth for ten years, the oak and beech of a very flow growth for near that time) to keep them warm and to encourage their growth. The practice is very wrong, and quite contrary to the prefent fyftem of thick planting. If the firs are planted at fix feet diftance, with an oak between, they will fmother the oaks in a few years; and if taken away fooner, they do not anfwer the end they were planted for. If the oaks are planted alone at three feet diftance, they will thrive much better, for they fuffer more from the cold when the firs are taken away, than any advantage they can receive from their warmth while they remain. When the firs are taken away your oaks ftand at fix feet diftance, which is too much, as the intention of planting thick is fruftrated, which is to prevent pruning and to keep the trees warm, both which are of the greateft confequence to plantations on poor land.

WHERE there is fo much rock, and indeed

deed no appearance of earth, there is no
poffibility of ftirring the earth with a
hack ; yet wc muft not defpair of raifing
trees and fhrubs, which is evident to be
feen in natural woods where trees and
fhurbs are feemingly growing out of the
ftone. No art can pretend to plant in fuch
places ; but nature fhews us what fhe can
do, and by following her dictates we may
accomplifh what has been thought impoffible.

IN all rocks there are openings and cavities, and by the moifture falling from
the higher parts of the rock into the bottom of the openings, there is fufficient
nourifhment to vegetate feeds ; and when
they are once in a growing ftate, the
young roots will find cavities and openings to pufh into, and alfo nourifhment
fufficient to make a tree, bufh, or fhrub.
It cannot be fo certain to get trees and
fhrubs to grow in rocks as in earth ; but
it may be depended on that many will
grow, and to a great fize. The only method is to drop feeds into the cavities.

THE beſt ſeaſon for dropping ſeeds into rocks is as ſoon as they are full ripe and dry ; but there are ſo many mice and other vermin about ſuch places in winter, not overſtocked with proviſion, that they de-ſtroy every thing within their reach. To remedy this as much as poſſible, the ſeeds may be ſo prepared as to be ſown or drop-ped in March with good ſucceſs.

THE preparation of the ſeeds for drop-ping amongſt rocks, and ſowing planta-tions on all kinds of poor land, will be treated of under that head.

IN order to drop the ſeeds amongſt rocks, let a man take a few of the follow-ing ſeeds, and drop three or four of a kind into each cavity, obſerving to drop the larger ſeeds into the deepeſt cavities, ſuch as acorns, beech-maſt, hornbeam, evergreen oaks, yews, mountain aſh, hollies, haws ; and, into the leſſer open-ings, broom, juniper, furze, birch, and wood elm.

THESE

THESE places fhould be gone over the following fpring, as there are many accidents to prevent the growing of the feeds. There will be no occafion to have any regard to drop the fame forts of feeds into the holes as they were dropped the fpring before ; for if both grow, it will be of no bad confequence, as we often fee two trees of different kinds growing in natural woods on bare rocks.

IT may feem ridiculous to drop feeds into rocks ; but it is evident, that many fine trees arc growing in fuch places, and it muft have been occafioned from feed accidentally dropping.   This is no more than following nature, and fhe has taught us what is to be done to cover fuch places as have been left naked and difagreeable.

IT may be objected, that there are many trees in natural woods, on rocks, that produce feed, and yet there are many bare places on the fame rocks. But it may be remarked, that many of the feed

falling

falling into one cavity may intice vermin to deftroy the whole, and many places where none fall. The fmall quantity that is dropped into each place is no great temptation to vermin, and the regular dropping prevents any being miffed.

I MUST here be underftood as fpeaking of fuch places where there is very little earth, and of bare rocks where there appears to be no earth, and where it feems almoft impoffible for any thing to grow, and where the fpade could be of no fervice. But I am certain, if the directions given are followed, a fine foreft will fucceed a barren mountain, which will be a great pleafure to the prefent poffeffor, a profit to pofterity, and an advantage to the kingdom in general.

HAVING given directions for planting the barren mountain where there is little foil, I fhall now take into confideration gravelly hills, heaths, and commons, where there is fo much earth that little holes may be made to plant and fow in.

MAKE the holes at three feet diftance as deep as the foil will allow, and one foot broad, if the ground is not very ftony at top. The readieft method is to pare off the furface with a paring fpade ; but if ftony, the broad end of the ax-hole will be the beft. If grafs, pare it off as thin as poffible, lay it afide, dig out the earth, and lay the pared furface into the bottom, laying in the earth in the form of a mole-hill, to remain until the feafon for planting and fowing. This work fhould be performed early in fummer for the rains to moiften the earth, which is always very dry when turned up, and there is no danger of weeds growing on fuch land. The earth being laid up round will have more advantage of being mellowed by the weather ; and when the planting feafon comes, there is nothing to do but to level the ground even with the furface and plant.

A MAN can make three hundred holes in a day, and two men may plant a thoufand, and do them well.

IF

If the furface be coarfe, benty grafs or
fhort heath, which is often the cafe in
fuch poor land, pare it off with the hoe
as thin as poffible, and throw it away.
The paring fpade fhould not be ufed, as
it takes too much of the fcanty foil: for
in fuch dry ground any thing that is light
keeps it open, and is of very bad confequence by making the ground lighter
and drier.

The holes being laid in this pofition,
will be in good condition to plant after
the firft autumn rains, for moft kinds of
deciduous trees. They fhould be planted
as foon as the leaf is decayed. They may
be planted in open weather the beginning of winter, but never in the fpring
on fuch dry ground.

Oak and larch are an exception to the
general rule; for although they are deciduous plants, their proper time of planting is late in fpring.

Oak

Oak thrives beft when it is removed in the fpring, juft before the bud begins to pufh; that is, about the beginning of March. This fhould always be obferved in the removing of them that have been trained in the nurfery for four or five years, as in a great meafure fuccefs depends on their being removed at a proper time. I have planted oaks of fix and eight feet high, the latter end of April with good fuccefs, but it was into a fine ftrong loamy foil.

The young plants of oaks that are intended for bare ground fhould be planted at the fame time with the trees from the nurfery; but if there is the appearance of a dry fpring, they muft be planted before the ground is dry, or they will infallibly perifh.

As it is beft to plant all kinds of trees by themfelves, it can be no inconvenience to leave the ground defigned for oaks until the proper feafon. This will be an advan-vantage; as the other forts of trees fhould be planted fooner, there will be more time to finifh planting.

IF there are any fmall crooked oaks in the nurfery, and feemingly good for little, prune their ftems, and plant them by themfelves; and after they have been planted two years, in any of the winter months cut them down an inch below the ground, and they will make fine ftraight fhoots next fummer. In the latter end of June, or beginning of July, let them be gone through, and all the fhoots but one of the ftraighteft and ftrongeft be fliped off with the finger and thumb, clofing the earth round the remaining plant; they will come eafily off at that feafon, but if they ftand much longer, we fhall be in danger of tearing the bark off the ftool, which would fpoil the tree. There muft be no knife made ufe of; for if cut, they will pufh many fhoots at every amputation, which would much injure the trees, as they muft be cut again, and

by

by that means will form a bunch round
the root of the tree, which would be very
detrimental, if not wholely deftroy it. It
will be neceffary the next fummer to go
over them at the fame feafon, and ftrip off
any fmall fhoots that may have fprung,
after which no further care will be wanted. This is foreign to what I propofed
treating of; but I have feen fuch oak
plants thrown away as good for nothing.
I have planted fuch, and treated them as
here directed, and have had a clump of
fine trees, ftraighter and finer than thofe
planted with their heads. It is my opinion, that all planted oaks which do not
thrive, if they were cut down as above,
would make fine trees. I have done fo
with fome, and they anfwered very well.

THE larch, although a deciduous tree,
fhould never be planted in winter, and in
moft autumns it is too late before it lofes
its leaf. The beft feafon then is the beginning of March, both for large and
fmall trees. They fhould be planted juft
before

before the buds begin to pufh. It is a
refinous tree, although not an evergreen,
and has fmall fibrous roots like all kinds
of firs and pines, whofe roots fhould never
be cut, unlefs they have been long out of
the ground.

None of thofe trees fhould be planted
when the wind is high, nor when the air
is frofty; their ftrong roots being hard
and brittle: the fmall roots, if dried by
froft or wind, never recover, and it is
from thofe fmall roots the tree is firft put
in a growing ftate: this often occafions
the lofs of the tree, as the large roots are
fo hard, that they feldom pufh until the
tree is growing. Young larches have only fmall roots, and require no cutting if
planted foon after taken up. If the fmall
roots are dried, it is of bad confequence,
and is the reafon they fo often mifcarry
when kept long out of the ground. All
kinds of refinous plants fhould have their
roots wrapped in wet mofs, if they are to
be carried to any diftance.

Vol. I.   B   All

ALL kinds of pines and firs fhould be planted early in autumn or in the fpring; the latter is preferable, as early in autumn the ground is generally too dry, and after the rains have fallen, it will be too late, as the frofts may be expected foon; and fhould they be very fevere, it will throw the young trees out of the ground.

THE Scotch firs are an exception to the general rule, for they may be planted with fafety from September to April; but in poor land that is hard and dry, it would be of great fervice to put fome grafs, ftubble, or any light ftuff round them, to keep out the froft, as they are fubject to be thrown out of the ground as the other kinds.

IT fhould be obferved, that firs ought never to be planted in fuch dry ground, as here treated of, when very dry; neither fhould they be planted when the ground is very wet. Two days after rain, it will be in good condition, that is,

in

in high ground; but there are fome very
poor lands that are fo flat, that the water has no way to run off. Such places
fhould be planted in fpring.

ALTHOUGH the ground in general be
very poor, yet in thofe places where the
water lies, it is richer, and generally of
a ftrongifh loam, although fhallow, and
a clayey bottom. If the water can be
drained off, fo that the ground remains
only wettifh, and not to ftand to cover
the furface, oak will thrive well in fuch
places, and grow fafter than in any other
ground. I have had them fhoot three
feet in one year, being the fecond after
planting, and very ftraight and ftrong,
although the whole ground was feveral
times covered with water the firft winter
after being planted.

WHEN fuch places are planted, the
holes muft not be made but as you plant;
for if the holes were to be made as before
directed, the bottoms would be full of

B 2   water,

water, the tree would ſtand in a quagmire, and it would be impoſſible to faſten
it. If the tree is planted ſix inches above
the level of the ſurface, it will be better,
and the ground made up, round the
tree, in the form of a round-topp'd muſh-
room.

In theſe places we muſt deviate from
the general rule for planting poor land.
The trees planted here muſt be ſuch as
have been trained in the nurſery for four
or five years, and of a pretty good growth.
The reaſon is, that the ſmall fibres of
young trees are eaſily rooted ; therefore,
ſuch trees as are planted in this ground,
muſt have all their ſmall roots pruned off.
The ſtrong roots will bear the moiſture,
and puſh freſh roots ſooner than in any
other ground. It will be abſolutely neceſſary to plant ſuch places in ſpring, and
it may be performed ten days later than
in any other grounds, for there will be
ſufficient moiſture for them all the ſummer.

THERE

THERE muſt be no places left for ſowing here; for the moiſture would rot the
ſeeds. Spruce-fir, and the ſwamp-pine
will thrive well in ſuch places; and the
plane tree will grow, if there is a foot
deep of ſoil. Every one of the poplar kind
will do very well; but I think they ſhould
all give place to the oak, which will thrive
to admiration.

POPLARS of all kinds have come into
great eſteem of late years, being found
very fit for every ſort of country buſineſs.
Several gentlemen have made chamber-
floors of them, which anſwer very well;
but they will not be of long duration.

As wood is ſcarce in many places, a
ſmall place might be allotted for a plantation of poplars; and as they are of a
quick growth, they will come to uſe in a
few years, and ſave better wood. They
will grow in any land that is not very hard
and dry; but as the intention of planting
them is, that they may ſoon become pro-

fita-

fitable, it is adviseable to plant them in a swampy rich soil.

The best method to propagate them is to cut them into truncheons of a yard long, and with an iron bar let them into the ground level with the surface, fastening the ground round them. They will push out many shoots the following summer, which should be all pulled off by the hand the end of June, or soon in July; only one of the straightest and strongest should be left to grow to a tree. There will be more shoots pushed out in the summer, which should also be pulled off as before (for they are very subject to push out suckers) after which they require no further care; for if they are planted at six feet distance, they will prune themselves, and grow very straight, and to a great size in thirty years.

Some sharpen them like a stake, and drive them down with a mallet, but that is not a good method; for if in driving,
                                    they

they are fcratched by ftones or roots, the bark decays on that fide, and often caufes a blemifh in the tree, after it has grown feveral years. When the bark is not broke, they pufh roots all along the truncheon to the very furface, and make much finer trees than thofe planted with roots; neither are they fubject to be blown up, which the others are.

IN plantations where under-wood is defigned, fome clumps of poplars will anfwer extremely well; if the ground is inclined to wet, they will be fit for cutting much fooner than any other wood; therefore they fhould be kept by themfelves.

THE fowing tree-feeds by the plow has been fome time in practice, and is certainly a good fcheme; but to ufe the plow in poor gravelly land, that is covered with fhort heath, and has never been in tillage, will anfwer no end; or if the field or common be rough with coarfe grafs, and little foil, it will coft ten times

B 4

the

the expence to bring it into order for
fowing or planting, than making the holes
for planting as before directed ; and the
trees will thrive much better than when
the coarfe grafs and heath are plowed in,
unlefs the ground is worked until they are
quite rotten, which would require a good
deal of labour and expence to no purpofe.

WHERE there is poor land, that has
growed corn for fome years, and is
defigned for planting, plowing will
be of great ufe, and a very profitable way
of working it; the field may, at a fmall
expence, be put into good condition,
which will promote the growth, fuccour
and encourage the trees for many years,
even until their own foliage becomes a
manure to them; and a crop of under-
wood my be expected.

PLOW the ground in autumn immedi-
ately as foon as the corn is off, and let it ly
all the winter to mellow ; plow it again
in

in the fpring, as foon as the ground is in condition; and a third plowing is neceſ-fary in order to fow turnip or rape feed, which ſhould be fown as early as poſ-ſible, and very thick. When it is grown fluſh, that is, juſt before the plants begin to ſhoot (for they will all ſhoot that are thus fown fo early and thick) feed them off with ſheep; as foon as they have ate them quite bare, plow the ground a fourth time; and if many weeds ſhould grow, which is hardly to be expected in poor gravelly land, plow it again juſt before winter. When it is dry, let it ly all the winter rough until fpring.

THE middle of April is the proper time for fowing the prepared feeds; but if they are fown without any preparation, they ſhould be fown the beginning of March. Early in the Spring there is al-ways a great hurry of bufineſs; it will, therefore, be a great relief to have a week more to perform the fame work; and, befides other advantages, this is a
very

very effential one for preferring the prepared feed.

Just before you intend to fow, plow
and harrow the ground ; and if the field
is fo large as to take feveral days work,
plow no more than you can finifh in a
day.

The ground being plowed and harrowed, mix beech-maft, afh-keys (the
afh-keys fhould be buried a year in a pit,
mixed with fifted coal-afhes, as they never
come up the firft year) fycamore, hornbeam, common black cherries, Spanifh
chefnut, maple, and acorns, and fow
them broad-caft all over the field ; then
plow the field with a very fhallow furrow three inches, and fcatter fome
birch and rough - leaved elm-feed all
over it.

The proper timber-trees for fuch land,
with underwood, are oak, beech, and
Spanifh chefnuts, which are all much of
a growth ;

a growth; and although this is not the proper foil for the oak, it will anfwer very well when fown. Notwithftanding there are all thofe kinds of feeds fown in the broad-caft way, and covered by the plow, it would not be amifs, at thirty feet diftance, to plant three of the timber-tree feeds in a triangle of fix inches di-ftance, and then there will be a certainty of having the trees, which are to grow to timber, at a regular diftance (which might be marked as the labour would not be much) and room for the underwood to grow fo as never to be over-topp'd by the timber-trees. The whole fhould then have a fingle ftroke with a bufh-harrow.

If any large weeds, fuch as thiftles, docks, or wild muftard, fhould grow, they fhould be pulled by the hand when young; as for fmall weeds, they will be of fervice in winter, to prevent the frofts throwing the feedlings out of the ground, and will not grow in fummer to be any ways detrimental to the plants; befides, they

they will keep out the drought, which
will be of great ufe in fuch ground.

THE trees defigned for timber fhould
have nothing of the tree kind grow nearer
them at firft than three feet; and the third
year after fowing, there fhould be only
one plant left in a place.

THE whole ground fhould be gone
over the third year, and all the plants that
are for underwood, and nearer than a
foot, fhould be drawn up, and the ground
faftened round the remaining plants.

As there will be a number of young
plants drawn, they fhould be planted in a
nurfery prepared for them in tolerable
good land, and trained for fome years to
plant in good ground.

ALL kinds of firs and pines muft not
have any place where underwood is intended; if allowed to grow fcattered
about the field, they fpread a vaft way,
and

and by the clofenefs of their branches will
deftroy every thing under them ; and if
they are cut down after they are grown to
a confiderable fize, they make a large gap.

THIS is the only method to get under-
wood in poor gravelly land. It will be
long after fowing, before it is fit for ufe ;
but after it has been cut the firft time, it
will grow better every cutting, and laft
many years.

GROUND thus managed with the plow
would make a fine wood, fown after the
following manner. Make the holes at
three feet diftance; put three or four feeds
in a hole, at four inches diftance, in form
of a fquare. To the weft, north-weft,
north, north-eaft, and eaft, make a fkirt-
ing of Scotch firs, and another of larches
in the fame afpects ; then beech, Spanifh
chefnuts, and acorns. The feedlings may
grow to the fecond year, when they
fhould be drawn, except two of the beft
plants. Then the third year, the beft
fhould be left, and the others taken away.

There muſt neither be ſpade nor fork uſed in taking up the ſupernumerary trees; for if the roots of the trees that are to remain are diſturbed, it will greatly hinder their growth. If a proper ſeaſon is obſerved, they will draw eaſily. After a very hard froſt, when the ground is thoroughly thawed, they will draw with the hand. When the plants are all drawn, faſten the ground round the plant that remains, with your foot. The plants taken up in this, and all ſown plantations on poor land, muſt be carried to the nurſery.

A poor cold clay, that has been in corn, and is of no great value for that purpoſe, will anſwer to make a plantation with underwood. It will not be very vigorous for ſome years; but after the roots have got good hold of the ground, it will ſhoot in faſt, and laſt many years.

Oak is the only timber-tree to be encouraged in ſuch a ſoil. Its progreſs will
be

be flow, but it makes fine wood. The acorns fhould be fown four in a place, at fix inches diftance from each other; and the places fhould not be nearer than thirty feet, to allow room for the under-wood.

Ash, poplars of all kinds, tree fallow, and birch are the proper underwood to be planted and fown in fuch land. The poplars and fallows, for planting in clay, fhould be cut eighteen inches long. As the clay is hard and dry, there will be little moifture to encourage their ftriking roots at a greater depth, therefore they fhould be let down, as before directed, in making plantations of them.

The ground fhould be plowed early in autumn, as deep as the plow can go, with a thin furrow, and ly all winter to mellow.

The poplars and fallows may be planted any time in winter, and the feeds fown in the middle of April. This land muft

not

not be plowed in the fpring, as it would be too ftiff for the feeds to vegetate in at firft.

THE acorns being planted at their proper diftances, as alfo the fallows and poplars, let afh, birch, and rough-leaved elm-feed be fown all over the field, and covered with a fingle ftroke of a bufh-harrow.

THE poplars and willows fhould be planted at eight feet diftance ; but none of them fhould be planted nearer the acorns that are to remain for timber-trees, than ten feet.

IN fuch a plantation it would be very beneficial to prune the oaks for ten years, which would make them have fine ftraight ftems, and be of great value. Being planted at fo great a diftance, there would not be a great number in a large field, and the labour and expence would be trifling. Where underwood is intended, the timber
ber

ber-trees fhould never be nearer, for if they
are, they will in a few years quite fmother
the under-wood, fo as to render it of little
value.

THERE is another advantage in pruning
the Oaks; their ftems being long, and
their heads fmall, the under-wood will
thrive all around them as near as it fhould
be allowed to grow; which will not be the
cafe if they are not pruned: they will
then have fhort ftems and very large heads,
and will deftroy the under-wood for thirty
feet round them.

IN countries fcarce of firing, and where
poles and rails are wanted, under-wood
will pay the proprietor triple more value
than the beft fields of corn, even allowing
for the expences of planting, fencing,
and the rent of the land from its being
planted to the firft cutting, after which
there is no labour but keeping up the
fences; fo that the profit will increafe, and
the Oaks for timber ftill remain a great
eftate to fucceeding generations.

CHAP. II.

On Sowing Tree-Seeds with Corn.

SOWING tree-feeds amongſt corn is an exceeding good method, for it prevents the vermin deſtroying the trees as they come out of the ground ; moſt trees bring up the feed with the firſt leaf, and all birds are fond of them, and often deſtroy whole fields as they come up. This method ſhould not be attempted on fields that have been in graſs, but ſuch as have been in corn for ſome time.

IF it were rich deep land, it would be a very good method to have a crop of underwood ; but bare poor land is our preſent ſubjeɛt, therefore I ſhall give proper directions for raiſing a field of good timbertrees on ſuch land.

THE

The ground fhould be plowed in autumn, and early in the fpring fown with
a thin crop of oats. If the feeds are dry,
and not prepared, they fhould be fown at
the fame time; but if prepared feeds are
ufed, which are much preferable, the
middle of April is the proper feafon.

The large feeds may be planted with a
fetting-ftick, not too fharp at the point, at
three feet diftance, three or four feeds in a
place, at four inches apart and two inches
deep.

The rough-leaved Elm, Larches, Scotch
Firs, Silver and Spruce Firs, and Pinafters
muft not be planted with the fetting-ftick,
but the ground hollowed out half an inch,
and three or four feeds dropped into each
place at four inches diftance, and preffed a
little down with the hand, then covered
with mould level.

The proper feeds to be fown, even in
poor land, amongft corn, are Oak, Beech,

C 2   Black

Black Cherry, Spaniſh Cheſnut, Hornbeam, Silver and Spruce Firs, the Scotch
Fir, Larches, and Pinaſter; and, if the
ground is inclined to ſand, Sycamore,
Horſe Cheſnut, and Limes may alſo be
planted.

ALTHOUGH the ground is not the propereſt for all the trees here mentioned, the
growing corn is of ſo great ſervice to them,
that they will thrive beyond all expectation.

THERE ſhould be no tree-ſeeds ſown
where the ſurface is not quite rotten. If
any gentleman chooſes to be at the expence of paring and burning fields or commons that are in graſs or ſhort heath, let
him ſow it with turnips, and let them be
eat off with ſheep early in autumn, then
plow it as deep as the ſoil will allow as
ſoon as it is clear of the turnips. In the
ſpring ſow it with a thin crop of oats, and
with the tree-ſeeds at the proper ſeaſon;
March

March for the dry feeds, and the middle
of April if they are prepared.

If the ground is free from large ftones
and tolerably even, fo as to be eafily plowed,
the expence will not be great, efpecially if
the feafon anfwers, the turnips will repay
the expences; there can be no expectations
of a great crop from fuch poor land. It is
for the advantage of the tree-feeds that I
recommend this method of working.

All the Pine and Fir kinds of Trees do
much better fown amongft corn than any
other way; and indeed it is the beft
method of fowing them, for the ftubble is
very beneficial to them, as it prevents, in
a great meafure, the froft throwing them
out of the ground, which all forts of Firs
and Pines are fubject to when fown otherwife.

All the different kinds of trees fhould
be fown feparate: the afpects from the
C 3   fouth-

south-east to the south-west should be sown first. A skirting of Scotch Firs, forty feet broad, should be sown if the plantation is large; then another of Larches of the same breadth; and if all the other sorts were sown in the same form, the different shades would add greatly to its beauty.

THE reason for sowing the Scotch Firs and Larches to the west, north, and east points is, that as they are very hardy, and quick growers, they will be a great shelter to the other trees, and keep them from the cold winds that blow from those quarters, which often bends them and makes them grow crooked.

WHEN the oats are cut, the labourers should be particularly careful not to tread on the trees. The oats should be cut high to prevent the seedling-plants from being topt; and the long stubble will keep the trees warm all the winter. The oats should be carried off the field by men

as

as foon as they are cut, for it will be impoffible to bring carts or horfes into the
field without deftroying many of the
young plants.

Trees fown amongft corn will grow
more the year they are fown, than they
will any other way in two, and be very
ftraight. I have had Oaks grow eight
inches from the feed in one year in very
poor land.

Unless there are thiftles, or fuch
ftrong-growing weeds (which fhould be
pulled up by the hand when young) there
will be no occafion to hoe or clean in fuch
grounds, for the ftubble and fmall weeds
will be of great fervice to keep out froft
and drought, which is of great importance
to the young trees on fuch dry ground.

The young plants fhould be thinned the
fecond and third years, as directed in the
other fown plantations ; after which they

C 4   will

will require no further trouble for many years.

It may be imagined that *sown* trees left at three feet diftance would be fo thick as never to grow to timber; in tolerably good foil it would be improper, but in poor gravelly and fandy land they grow very flow for feveral years, and never begin to make any great progrefs until they fhelter and keep one another warm; they alfo grow much ftiffer and ftronger in the ftem, and their fide-branches do not extend near fo faft as in good land; therefore their being fown thick is a very great advantage, and the only method to make them thrive.

There is nothing further to be done until the plantation becomes quite a thicket, then every other tree muft be ftub-felled, and they will be fit for many ufes.

The trees to be taken away (as they will not be drawn weak on fuch ground) may

may feem very good plants for tranfplanting; but that muft not be attempted, as
it will be impoffible to take them up with
any thing like a root for planting, without
greatly hurting the roots of the trees that
are to remain on the ground; it would
greatly retard their growth, and the trees
taken up would have little chance of
growing, being taken out of fo clofe a
fituation; and it would be impoffible to
avoid cutting thefe roots very fhort, as the
trees are only three feet diftance, with their
roots fpreading all over the ground.

ALL the plantations here treated of muft
be thinned in the fame manner when they
are grown up; then the trees will ftand at
fix feet diftance, which will be fufficient
room for them to grow to timber. After
fome trees are taken away, and the remaining ones are at fix feet diftance, their
fide-branches will ftill near meet, and be
fufficiently warm; and they will decay as
the trees advance in height, and by their
clofenefs will prune themfelves.

It

IT will be a great temptation to lovers of trees and planting, who have plantations going on, to thin their fown plantations. When the trees are about fix feet high they are then of a proper fize and age, and would be very fit for tranfplanting into any ground that is tolerably good; but then, the plantations they were taken from being on poor land, they would never thrive, but ftunt and grow crooked bufhes, as may be feen in natural woods on poor ground. Where the trees are thin they make no progrefs, whereas where they are very thick in the fame wood and on the fame foil they are very ftraight and tall; for which reafon no temptation muft prevail to thin them at that age.

WHERE under-wood is of great value, and where there have been large plantations made on poor land, when they come to be about fixteen feet high, if there was a hundred yards cut clean off by the ground, then a clump of trees to ftand of the fame fize, and fo on all over the plantation, this would

would be profitable to the prefent proprietor and advantageous to the country ; and
as the trees ftand in large clumps they will
thrive very well, and the under-wood will
grow very faft after the firft cutting.

CHAP.

C H A P. III.

On Planting Moors and Commons covered with long Heath.

THERE are large hills, moors, and commons covered with long heath; such places are of more value than is generally imagined. The heath which grows on them plainly fhews the ground is fertile, and if they had been planted fome years fince would have been fine forefts, and not one fprig of heath to be feen. It is the beft foil for planting in England, and will bring trees to as great perfection as our beft land, and with fmall expence.

THIS foil, in appearance, is a light, black peat earth ; but it is of a far fuperior nature, for on that black, loofe, moffy earth grows little heath, only a ftrong benty grafs, and a little foft moffy earth below the grafs, then a hard kind of peat, which

which is very barren; whereas that on which long heath grows is of a great depth, and of a fine moift clammy nature. On this foil moft of all the kinds of foreft-trees known in England will grow, even all the kinds of poplars, which fhews that it is of a very fine moift nature.

THE planting or fowing in fuch ground is much eafier performed than in the beft land: there is nothing further requifite than to pull up the heath about a foot's breadth, and if you fow with the large tree-feeds, fuch as Acorns, Spanifh Chef-nuts, or Beech-Maft, they may be planted with the fetting-ftick (as the pulling up of the heath has loofened the ground fuffi-ently) two inches deep; four or five feeds in a hole, three inches apart.

WHERE the fmaller feeds are fown, fuch as the Silver and Spruce Fir, the Scotch Fir, the Pinafter, Weymouth Pine, and Larch Tree, after the heath is pulled up, chop the ground with a fpade, and drop

four

four or five feeds into the place, at three or four inches diftance, then gently tread it with the foot; and nothing further is required in fowing on ground where long heath grows.

Such ground requires lefs labour to plant than any other: there is no occafion to make the holes until they are juft going to be planted, and they need not be any larger than juft to receive the tree roots.

The heath is of great advantage to the young trees, and makes them thrive exceedingly if it is higher than the trees planted, which I would advife. If the heath is eighteen inches or two feet high, the trees fhould be a foot or thereabouts; the heath will then keep them warm, and will protect them from all winds until they have got good roots; fo there will be no occafion to plant fo thick as on the poor bare land.

Six

Six feet will be a proper diftance; the trees will grow very faft, and by the time they have got a foot or two above the heath their branches will near meet, and as the trees advance the heath will decay. The trees growing clofer will prune themfelves, and there will be no further care requifite.

If the ground is a flat moor, it fhould be planted to the weft, northern, and eaft afpects with Scotch Firs and Larches, as before directed upon poor land, for fhelter.

If a hill or rifing ground, which is generally the cafe, the higheft ground to the weft and north fhould alfo be planted with Scotch Firs and Larches, to fhelter the other trees from the north winds; but the plantation of them fhould be fmall, as it would be bad policy to ufe much of this ground for any trees but the Oak, which thrive in fuch foil better than in any other.

If

IF such rising ground is in sight of a gentleman's house, there may be some clumps of the different kinds of Firs and Pines planted for ornament, as they are very beautiful at a distance in the winter. The Silver and Spruce Firs will grow to an immense height in such soil.

ALTHOUGH I have given directions for sowing and planting the different kinds of forest-trees on such ground, it is to shew that it may be planted and sown with such where gentlemen are inclined so to do; but as the Oak is by much the finest and most valuable wood, it should have the preference, especially on ground so proper for its growth.

I MADE three different plantations a-mongst long heath (where the ground answered the description given) in the same year. The first was sown with Acorns and Scotch Fir-seed; the second was planted with seedling Oaks a foot high, two years old, and Scotch Firs from the seed-bed of

that

that year's fowing, but were thinned to two inches in the feed-bed, and were fine ftiff plants; the third was planted with feedling oaks, three years old (eighteen inches high) but had not been removed from the feed-bed. The heath was near two feet high and very ftrong.

In the firft, fown with Acorns and Scotch Fir-feed, many of them came up very well; but the mice, which were very plentiful amongft the heath, deftroyed many of the acorns, fo that many of the holes were quite empty, and many had only one tree. Next winter many of the young fhoots of the Oaks were cropped by the hares. They were to make good for three years before all the holes were full, which was a good deal of trouble and expence.

The growth of the plantation was very irregular; the Firs in fome places fpreading over the Oaks, and where they had been fown the laft time, to make up the deficiencies, not fo high as the heath.

THE young shoots of the Oaks were weak, being drawn, but very straight, and when they got above the heath they were slender for some years. So that I think planting good young plants is better than sowing acorns in such ground.

THE heath had no bad effect upon the sown Firs, they grew fast and strong, and there were few of them destroyed by the mice; there were many to take away, and those plants were much better than those sown in seed-beds according to the common method, and fit to plant in any ground; therefore if a plantation of them were intended and sown alone, they would answer full as well as planting amongst heath.

WHERE the seedling Oaks and Firs were sown together, the Firs did not get the better of them for some years, but their side-branches at last spread so, that it became absolutely necessary to take them away. The whole were accordingly taken away.

away, and replaced with Oaks of fixteen inches high (two years old) from the feed-bed. They all throve very well, and a few years after the plantation was pretty equal, only the planted Oaks were ftiffer than the fown, and not fo tall. No Firs fhould be fown nor planted with Oaks, let the ground be good or bad.

In the fecond plantation, both Oaks and Firs did very well; but in fix years the fide-branches of the Firs over-topp'd the Oaks; and the feventh year there was a neceffity of taking all the Firs away, which was done, and Oaks of three feet high, that had been removed from the feed-bed into the nurfery two years, planted in the room of the Firs.

They all grew very well, but the firft-planted oaks were the fineft trees ten years after, which is a good reafon for making all plantations for timber of fmall trees: for unlefs it be for pleafure, where large trees are planted for ornamenting pleafure

D 2   grounds,

grounds, without any regard to timber, the small-planted trees will in time make the finest woods, and in less time than those that are planted large.

THE third plantation, planted with Oaks only, of three years old, eighteen inches high, and had never been removed from the seed-bed, grew extraordinaay well, and was the finest of all the three. There was no further trouble with it.

THE heath decays as the tree advances in height; and as they grow thick, the under-branches decay, and they have in general fine straight stems.

THERE was on the same hill I planted on a plantation of Scotch Firs, which had been planted sixty years. At the time the Firs were planted, such high grounds were thought to be fit for no other kind of wood. They were fine trees and many of them were cut for use: the wood

was

was pretty good, but not come to its full growth. There was not one fprig of heath on all the ground the Firs were planted on, and if it had been any other kind of wood but Firs, there would have been good feeding for cattle and fheep, and good fhelter.

ALTHOUGH, in all plantations where fowing the feeds is recommended, I have given directions for fowing all the kinds of Fir and Pine-feeds, yet the fowing them on the place they are to remain in, on poor ground, is not the beft method (unlefs amongft corn or long heath) as they are fo apt to be thrown out of the ground, efpecially on poor land, where they grow very little the firft year, and are very fmall; and the ground being naturally very loofe, they are froze in winter below the roots, and many of them thrown out of the ground, and fome of them are raifed fo as to ftand like a fpider, with a very fmall part of their roots in the earth.

THOSE

THOSE that ſtand ſo will grow, but their being raiſed ſo retards their growth much, and they look red and ſhoot little until their roots have got hold, which will not be for a year or two. A little ſoft moſs ſpread over the holes when they are ſown is of great uſe to prevent the froſt's penetrating the ground below the ſeedling's roots; but that is attended with ſome inconveniency, for it muſt be at leaſt an inch thick, or it will be of no ſervice; and then it encourages grubs and other vermin to lodge under it, and they very often deſtroy the young plants as they come out of the ground. The beſt method for all the Fir kinds is to ſow them and plant them from the ſeed-bed.

THE compoſt for the ſeed-beds, and the method of ſowing and managing them, ſo as to make them fit for planting out the firſt year, will be fully explained under the article of ſowing all kinds of tree-ſeeds.

I₁

IF there was any mofs fpread over the holes and places where the Fir and Pine-feeds were fown, it muft not be removed when the fupernumerary trees are taken up, unlefs the ftem of the plant that is to ftand be earthed up to the top; for it will be very tender by growing through the mofs, and would be in great danger of perifhing by the cold winds in fpring. Earthing up the ftem is more advantageous to the tree, and will greatly encourage its growth, but will be attended with a good deal of labour, as the ground fhould be a foot level round the plant. If the mofs be left, the ftem will harden as the mofs decays, and the plant will thrive very well.

THE difference between planting and fowing Firs is trifling, for both in rocky and poor land the holes and places are made the fame for planting as fowing; and it will take very near as much time to fow a hole as to plant a fingle tree into it; and the advantage is very great in favour

D 4

of planting, for the many reafons before given.

A MAN may, with care, plant four hundred in a day, where the holes are made, and do them well. As they are planted, it would be of great fervice to them to lay fome long grafs, heath, or mofs round each ; it will not be much trouble, as the parings which come off when the holes were made will be near at hand. It will keep out the drought in fummer, and prevent the froft from loofening the mould in winter; for although the froft is not fo detrimental to the planted tree as to the feedling plants, the tree will thrive much better when the mould is not fo light about it.

FIRS and Pines are in general planted all the winter months, efpecially the Scotch Fir, and very often with good fuccefs ; yet it is not the beft method in poor ground, for the reafons before given. Early in autumn is the beft feafon for planting them

6   in

in hard dry ground, provided the holes were made early and had fome rain.

It is the nature of the Fir to pufh roots immediately after they are planted, if the weather is free from froft for fome time. It will be a great advantage to trees of any kind to be well fettled on dry grounds before winter, but there is none of them that get frefh roots but the Fir kinds. Firs may be planted to the end of April, but at that feafon they fhould not be long out of the ground.

Where there are large plantations of Scotch Firs that come to maturity, they bear feed in great plenty when about thirty years old, If there is any wafte ground joining the Fir wood that is covered with heath four or five inches long, if it was inclofed, and kept from cattle and fheep, in a few years it would be very full of fine young plants, and will ferve for a nurfery for feveral years, and ftill there may be fufficient trees left to grow into a wood.

WHEN the trees are grown to eight and ten feet high, the young plants that fpring yearly will not be fit for planting, as they will be drawn weak: and although fome of them fhould be ftiff fhort plants, they will be in danger of perifhing when taken out of fuch clofe places. So if there are many of thefe trees wanted, it will be the eafieft way of propagating them to take up yearly all the young trees that are fit, until there is a fufficiency got; then they may be allowed to grow into a thick wood without any further care; and as they grow fit for ufe may be thinned as wanted.

THERE will be a great many to cut out, but there is no need of thinning them to any regularity or fet diftance, for the thicker they grow they make the finer wood. A great many will be over-topp'd and decay, but that is not minded in countries where thofe woods all grow from felf-fowing, and are the fineft timber.

IF

IF the heath adjoining the Fir wood be
long, when inclofed it will anfwer to make
a wood, but will be of very little ufe as
a nurfery, as the length of the heath will
draw up the trees weak, and they will be
void of branches on the fides until they
are grown above the heath, fo will have
long weak ftems, and will not be fit to
plant. They may be allowed to grow
from the firft coming up; they will grow
very faft, and foon be fit for ufe.

BUT fuch grounds as produce long
heath, may be better employed than with
Fir-trees, as has been treated of in planting
ground covered with long heath. I have
heard of fowing Fir-feeds broad-caft
amongft heath, as corn, without any further trouble, and that it has fucceeded very
well.

THIS is fomewhat fimilar to what has
been juft now treated of, and I make no
doubt the hint has been taken from that,
or by feeing the feeds of trees come up in
the

the grounds at a diftance from the trees
that bear feed.

But there is a great difference between
ground around, and even at a diftance from,
feed-bearing trees being full of young
plants, and fowing in ground where there
is no fupply. Round a feed-bearing tree
there will be plenty of young trees if cattle
are kept off; and where there are vacancies
it may be replenifhed next year.

Whereas the feed that is fown at
random there muft many perifh, and many
more be deftroyed by vermin ; and as there
is no fpupply, there may be many places
empty ; and as there is not the advantage of
trees full of feed to make up the deficiencies,
fo there muft be a want, if not made good
by fowing frefh feeds, which would be troublefome and expenfive, as perhaps there
would need repairs for years ; fo that the
expences will be greater than following the
directions given for fowing Fir-feeds a-
mongft long heath, which is little trouble,
and fo eafy that it may be performed by
any labourer at a very fmall expence.

A LITTLE quantity of feed will fow a great deal of ground. One man will fow and make the places for feveral hundreds in a day; and there is almoft a certainty of fuccefs : and what is more advantageous, if four or five good feeds are put into each place (as directed for fowing Fir-feeds amongft long heath) at three inches diftance, there is a probability of having three or four good plants to fpare in each hole for making more plantations.

WHAT has been faid of the fuccefs of the inclofed wafte has been done frequently. There are many objections againft fowing the feeds at random. It were to be wifhed that fome public-fpirited gentleman would try the experiment, not only with the Fir, but feveral other kinds of tree-feeds ; it would be a great advantage if it an-fwered.

THE preparation of tree-feeds is of great advantage, as it will in a great meafure prevent their being deftroyed by vermin, which

which is one principal objection for not
fowing all the poor land inſtead of planting; for ſown ſeeds grow much better
than planted trees, eſpecially the larger
ſorts.

THE beſt ſeaſon for ſowing all kinds of
Acorns, Beech-maſt, Cheſnuts, &c. would
be the autumn, as ſoon as they are quite
dry, were it not that they have to lay all
winter in the ground before they vegetate;
and the mice, who are very dextrous in
finding them, will often deſtroy a great
part of them; and frequently what they
do not eat they will collect from different
places into holes of their own making,
leaving many places in the field without a
ſeed, as I have often found to my great
diſappointment.

To prevent as much as poſſible theſe
inconveniencies, the prepared ſeeds, with
equal ſucceſs, are ſown late in the ſpring,
and are a very little time in the ground
before they come up, and after they have
begun

begun to vegetate, the mice are not fo fond of them.

THERE is alfo another great misfortune that tree-feeds fown in autumn are liable to, that is, if there fhould come a few weeks of fine weather in the end of January or the beginning of February, it will caufe the feeds to come up early, before the hard frofts are over, which is the deftruction of many of the young plants, and fometimes of the whole.

THIS does not, however, often happen, as there is feldom fuch fine weather at that feafon; but it is a misfortune feeds fown in autumn are rather liable to, which the prepared feeds prevent, the danger being over before they are fown.

THE fmall birds are great enemies to all the tree-feeds that are fmall, the Fir and Pine kinds in particular, and which bring up the feed with the firft leaf; but as there are few of them on uncultivated heaths
and

and commons, where there are no trees, they may eafily be deftroyed.

THE crows are the moft mifchievous, and will deftroy a large plantation in a fhort time ; for as foon as they find a field or common, fown with tree-feeds, they will root with their bills, and fcratch with their feet, until they get at the feed, although two inches deep in the ground, efpecially Acorns and Beech-maft, which they are very fond of.

THE beft method I could ever find was to fhoot fome of them, pull off fome of the feathers, fcatter them about, open the crop, put fome gunpowder on the infide, and drop the carcafes about the field : this will frighten them for fome time, and as the prepared feeds are foon out of the ground, it will be a great means to fave them from thofe enemies, who are not fo fond of them after fprouting.

THE feeds of Acorns, Beech-maft, and

all kinds of nuts fhould be kept in dry fand all winter, in a place that is not damp, for the dampnefs would make them fhoot too early in the fpring for fowing; nor in a place where there is any heat, for that would make them pine, but in a dry airy place. The Sycamore, rough-leafed Elm, and all the kinds of Firs and Pines, fhould be kept dry without any fand.

THE rough-leafed Elm is generally fown as foon as dry after it is ripe, which frequently comes up the fame feafon, but fometimes not till next fpring, and then for the moft part fo early, that many of them are cut off by the froft. If this often happens to them in warm nurferies, they would ftand a great chance of being all deftroyed if fown in fields or commons of poor land, as it is fo late in the feafon before the Elm-feed is ripe; for even thofe that come up would be fo fmall and weak that they would be all thrown out of the ground; and thofe that did not come up would perifh by wet and cold.

VOL. I.   E   BUT

But if the feeds are thoroughly dried, and kept from damp, they would do very well to be prepared and fown in fpring, at the fame time the other tree-feeds are, which is of great ufe, for they are very fit for many places on poor land, grow much better from the feed than when planted, and is a profitable good wood.

The preparation of tree-feeds, fo as to have them vegetate before they are fown, is what has never been practifed nor known, and is of the greateft ufe to fave the feeds from vermin and froft in fpring, and from being injured by the frofts in the winter following.

As to the fowing feeds in fpring, that has already been fhewn, and as they have all the advantages of thofe fown in autumn, and liable to none of their difafters, it is a great improvement; for as the fevere frofts in fpring are over before they are fown, and they, by being prepared, are near as forward as if they had been fown

3

in

in autumn, they have time to grow to a good fize, and to have their roots well fixed in the ground, and are not liable to be thrown out of the ground by froft the next winter.

THE preparing and vegetating tree-feeds before they are fown is only following nature; for the feeds that are blown by the wind from the trees into different places, thofe that grow are laid up by chance in fecurity, and vegetate by the warmth and moifture of the places they lay in fooner than any that are fown in the common way in the fpring, although fown as early as the weather will allow; and it is only doing what is every year practifed, although it has never been thought of.

HAWTHORN and Holly-berries are buried in autumn as foon as they are pulled, and lay a year, and then are fit to be fown. The pulp of thofe berries is hard and dry, and takes a long time to rot, and until

E 2   that

that is quite rotten there can be no vegetation; and if no art is ufed they take a
year; but as foon as they are pulled, if
they were mixed with frefh grains, and
turned over every three or four days for
one month, and then lay all winter covered,
fo that no froft can come at them, and
turned over fometimes to prevent mouldinefs, which would deftroy them, they
may then be fown in the fpring with the
fame fuccefs as if they had been buried a
whole year.

THE only danger in having tree-feeds
fprouted before they are fown, is having
their fprouts rubbed off in fowing; but of
this, unlefs the perfons employed are ex-
traordinarily heedlefs, there is no danger.

THE proper time for fowing Hawthorn
and Holly-berries, that have been buried
a year, is the beginning of March; but
by the hurry of other fpring bufinefs, it
may be fometimes the middle of April
before they can be fown; by that time they
are

are fprouted a good length.   I never found
it any detriment to their growing, for it
has happened to me feveral times.

I MENTION the Hawthorn, as every
one is acquainted with the nature of its
management, and knows it will grow very
well when fprouted before fown.   If the
tree-feeds are managed as directed, they
will not be fo much fprouted as the Haw-
thorns, and fo will not be in any danger,
but will anfwer every thing that has been
faid of them, for there is nothing advanced
but what I have practifed often with good
fuccefs.

THIS method prevents many of the mif-
chances that fowing tree-feeds are liable to;
for as feeds fown on hills, commons, and
fields, ought to be done with great care,
fo it would be very troublefome to have
the whole to, go over for a year or two to
make up deficiencies; befides.the irregula-
rity in their growth.

THE compoſition for preparing the ſeeds to vegetate is ſoft pit ſand (ſharp ſand will not anſwer) and freſh grains from the brew-houſe; the ſand ſhould be got in ſummer, and made very dry, and laid by until wanted; the grains muſt be freſh from the brewhouſe, but muſt be ſpread, turned, and dried until they are juſt of a clammy moiſture; then they ſhould be mixed with dry ſand in equal parts, until there is a ſufficiency for the quantity of ſeed to be prepared.

AFTER the ſand and grains are mixed, they muſt be rubbed between the hands, and laid in a heap for four days, then turned over every day for a week, then let them lay four days in the heap; and if then there is no mouldineſs, but a fine clammy moiſture, it is fit for uſe; but if there is the leaſt appearance of mould, it muſt be turned twice a day for three days, and then lay four days more; and if then there is no mould it may be uſed without danger, for here is nothing
that

that can be of the leaſt hurt but the mouldineſs.

WHEN the compoſition is ready, have
ſome boxes of different ſizes, according to
the quantity of ſeeds; if the quantities
are ſmall, garden-pots will do as well.

FOR the large ſeeds, ſuch as Acorns,
Beech-maſt, &c. lay an inch of the compoſition at the bottom of the box or pot,
then a layer of ſeeds; fill up all the vacancies quite level, then a layer of ſeeds; and
ſo on till the whole be finiſhed. The Fir-
ſeeds may be mixed with the hand and
laid in the box, firſt laying half an inch
of the compoſition in all the boxes or pots
you lay ſeeds in.

THE Elm-ſeed, Sycamore, or any other
ſmall ſoft ſeeds, ſhould be rubbed between
the hands with the compoſition, and ſo
laid into the box, laying half an inch of
the ſame ſtuff over all the tops of the pots
or boxes, and place them in a dry place,
E 4 where

where no wet can come at them, and if it were where they could have the benefit of the fun it would be better, but if that cannot be conveniently done, they will do very well without : if they have the fun, the boxes or pots fhould be turned every week, that each fide may have the fame advantage.

THE large feeds, fuch as Chefnuts, Acorns, &c. fhould be put into the compofition the middle of February ; the Fir and Pine-feeds, the beginning of March ; and the Elm and fuch other foft feeds, the middle of March : they will all be ready for fowing the middle of April. Great care muft be taken to keep them from wet and mice ; the wet would caufe them to mould and entirely fpoil them, and if mice were to get into the boxes they would deftroy the feeds in a fhort time.

IF the holes were made before winter, as has been directed, they will be in fine order for fowing the middle of April ; and

as

as at that feafon the fun has great force, and the holes being moift and mellow, the feeds will be above ground in a little time.

WHEN the feeds are to be taken from the pots and boxes, great care muft be taken not to rub off the fprouts. The beft way to carry them into the field for fow-ing is in little boxes with a handle; and as a fmall quantity will fow a great deal of ground, there fhould not be too many taken out at a time ; and it will be necef-fary to take fome of the compofition in the box with the feeds, to prevent the young fhoots being hurt by the fun and air, which would greatly damage them.

THIS work fhould never be done when the wind is high, nor when the air is frofty; a calm dull day is the beft; but if the fun fhines, if not frofty, it will do very well.

IF the heath, fields, or commons are planted

planted or fown alternately, the planting
being before finifhed, there is only the
holes left to be fown, which fhould be
the fame fort as the trees planted; for I
would advife by all means to keep the trees
in feparate clumps.

THE Chefnuts and all the other large
feeds may be planted with a fetting-ftick,
which fhould be a little thicker than the
feeds to be planted, that there may be
fome loofe mould lay round the feed; and
it fhould not be fmall at the point, as that
would caufe a vacancy between the feed
and the bottom of the hole. The holes
fhould be made two inches deep, and the
feed dropped into them; there fhould
be four feeds planted in every place, at
three inches diftance from each other. All
the holes, after the feeds are dropped in,
fhould be filled with loofe mould.

FOR the fmaller feeds, fuch as the Pines,
Firs, Elms, and other fmall foft feeds,
there fhould be four places made with the
hand,

hand, half an inch deep, a feed dropped into each, and covered level with loofe mould. A man may perform a great deal of this work in a day, and there is not the leaft doubt of fuccefs.

The propagation of trees to plant in poor ground has never been made a different article from raifing trees in general; but it is very different, and very eafy. In raifing trees, they are moved from the feed-bed, the roots dreffed, and planted in the nurfery, that the tap-root may be deftroyed and made fit for moving, that the roots may fpread horizontally, and not go too deep into the ground; but in poor land they are to be planted from the feed-bed, and the tap-root is to be preferved, (which feems a contradiction) for it is on its length and ftrength that the chief part depends; for without it there will be little hope of fuccefs in planting on poor gravel, in heaths, commons, and rocky places, where there is little earth.

Prepare

PREPARE some beds of good earth, and add to them a large quantity of sand; let them be well worked over three or four times, so that the mould and sand be well mixed a foot deep at least. It would be best to do this in the beginning of winter; and let them lay all the winter in small ridges, that the frost may mellow them, and that the sand and the mould may be well incorporated. Early in the spring dig them over and lay them flat; then, just before you intend to sow them, dig them over again. This seems a great deal of labour, but the success of the plantations, planted from trees here raised, will make it all well bestowed.

AT the same time there should be a good heap of the same sort of mould and sand that the beds are made of, as near as can be to them; and, if it can be had, a third part of black mould, from old woods, where sticks and leaves have rotted for some time, which should be all well mixed, and turned over several times to mellow.

This is to cover the feed-beds after they are fown ; for taking the mould out of the alleys to cover the beds (as is the common method) is not right ; for it is fo trod in fowing the feeds, that it is difficult to break it to cover the feed ; befides it makes the alleys fo deep that the fides of the beds moulder down, and many of the trees are loft, at leaft they ftand fo dry that they make no progrefs.

The beds being ready, they fhould be fown with the prepared feeds the middle of April, as they are the propereft, being forwarded in their growth by being vegetated before they are fown, fo will be fine plants in autumn, when they are to be planted in the fields where they are to remain.

The feeds may be brought in the boxes or pots they were laid in. The fmall feeds may be fown all over the beds, compofition and feeds together, but not too thick, and covered half an inch with the mould

mould that was thrown in a heap for that purpofe.

THE large feeds, fuch as Acorns, Chef-nuts, &c. may be taken out of the boxes, and placed on the bed with the hand, at three inches diſtance, and covered with the fame mould as the others, two inches thick: they will require no further trouble, but to keep them clear of weeds until they are taken up for planting. As the feeds were prepared, there ſhould none be fown but what have fprouted, that there may be no vacancies in the beds, if the mice can be kept away.

As foon as the fmall-feeded plants are come into the third leaf, they muſt not be fuffered to ſtand nearer to one another than three inches, that they may grow ſtrong and ſtiff, and have a free air in the feed-bed, which is of great utility to them when they change their quarters to a poor cold heath or common.

THE

THE reafon of their growing to the third leaf before they are thinned, is that they may be of ufe, and by ftanding a year in the beds they are planted in, will be as fit for planting in poor land as thofe in the feed-beds. It may be imagined that at that age they will be too fmall and tender, and will not bear tranfplanting; but they will, and grow very well, if carefully planted.

IF there is occafion for more trees than can ftand in the feed-beds, they fhould be planted; and there is this advantage, that as they muft ftand another year in the beds they are planted in, before they will be fit to remove into the commons where they are to remain for good, fo that there will be a provifion of trees for two years, equally good, from the fame fowing; but if they are not planted, they muft be pulled out of the feed-bed, for there they muft not ftand clofer than three inches.

IF the young plants, thinned from
amongft

amongſt thoſe ſown, are intended to be
ſaved, there muſt be ſome ſpare beds, the
ſame as the ſeeds were ſown in ; and the
evening before the ſeed-beds are to be
thinned, they ſhould be well watered, to
make thoſe that are to be drawn come up
eaſy. There will be ſome that will break
in drawing, but that muſt not be regarded;
for it would be quite wrong to looſen them,
as it would damage the plants that ſtand
in the bed more than the value of thoſe
that will be broke.

DRAW only a few at a time, make a
hole with a ſetting-ſtick, and let them
down the whole length of their ſingle
ſtraight root (for at that age they have no
fibres) and cloſe the earth gently to them,
for their ſtems are very tender.

GREAT care muſt be taken to let down
their root ſtraight its full length, or they
will not be fit for planting on poor land ;
but will anſwer very well for plantations
on tolerable good land.

AFTER

AFTER they are planted they fhould have a little water every other day for eight or ten days, but they muft have very little at firft; for as their ftems are tender much water would rot them. If the weather is clear, they fhould be fhaded for three or four days, and then they require no further trouble. I have had very good fuccefs feveral years in this very way, and never allowed feedlings to grow too thick in the feed-bed, which is frequently practifed: the bad confequence attending fuch management is too obvious to want any explanation.

THERE is one thing I cannot omit taking notice of. It is the general opinion that all trees fhould be raifed on a ground fimilar to that they are intended to be planted in; this is certainly wrong; and I dare fay that moft, if not all, the nurfery-men in England will join in the fame opinion.

THE whole of what has been treated of

VOL. I.   F   is

is planting on poor land, on which it is very certain a wood may be raifed both by planting and fowing; but it would be an odd attempt to make a nurfery to tranf-plant from on fuch ground. The feeds, when fown in fuch places, find nourifh-ment for their fmall roots to fupply their little heads, whofe progrefs is flow until they gather ftrength, and when they get to three and four feet high, their own warm-nefs makes them grow fafter than could be expected; and in the holes that are fown, the plants that are to fpare in them are not fit for planting in fuch ground; they have little heads and fmall fibrous tufty roots, and would be all thrown out of the ground in winter, at leaft fo loofened as never to make a tree.

If an animal was to be only half fed, from its firft having life, for one year, I believe that fuch an animal would never grow to be of a large fize of its kind, if afterwards it was put into better keeping; but fuppofe it was put to harder fare; I

believe

believe it would make a poor figure. If this fame animal had been moderately fed for one year, and then put it into worfe feeding, it would have been a better beaft.

THUS it is with trees; if they are fown in fuch poor hard dry land, they are hide-bound from the beginning, and it would be fome years before they would recover, were they tranfplanted into good ground. What then muft be their cafe when tranf-planted into the fame fort of ground, per-haps into a colder fituation?

I HAVE been longer on this point than I intended, becaufe it is often recom-mended that the nurfery be as near as poffible to the ground the trees are to be planted in.

ALL public nurferies in the kingdom, that are of fame, are on fine light good land; and many years experience fhews that all kinds of trees and plants bought from them thrive very well, although for

F 2   the

the moſt part planted in much worſe
ground than where they were raiſed.

IT would not be right to raiſe trees for
planting in poor land on ground made very
rich with dung; but it would be much
worſe to pretend to raiſe them on a very
poor ſoil. If the beds for ſowing the ſeeds
are made as here directed, they will produce
fine plants for the purpoſe, and there will
be no doubt of their ſucceſs when planted
in the very worſt ground.

THE management and ſowing the ſeeds,
as alſo the raiſing of the trees, for planting
poor land, are all fully explained.

WE now proceed to the planting, which
may be either alternate, or wholly detached.
I would recommend alternate planting and
ſowing, unleſs where it is otherwiſe
directed; that is, all the Fir and Pine
kinds ſhould be *all* ſown or *all* planted;
but planting them is much the beſt, as it
is almoſt impoſſible to ſave the young
feedling

feedling plants from being hove out of the ground on poor land in winter; fo it would be better to let the Firs, Pines, and Larches be two years old before they are planted, but they muft not be moved from the feed-bed, but thinned as before directed.

THEIR roots are fmall and tufty, and eafily managed, and may be fpread about when planted without cutting any of them, which will caufe them to get good hold of the ground, and be in no danger of being thrown out of the ground by froft.

ALL the deciduous trees, but the Oak and Larch (which are much better to plant in fpring) may be planted from the decay of the leaf, in any of the winter months when the weather, is mild, to the latter end of February, after which it will not be proper to plant in poor land and in hard dry ground; for if the fpring fhould prove dry, which it often does, thofe

F 3   planted

planted later would be in great danger of being loft.

ALL the trees muft be planted from the feed-bed of the fame year's growth; but thofe that were thinned from the feed-beds and planted out, will be fo much retarded by being moved, that they will not be fit for planting till the fucceeding winter. Thofe that were fown in the feed-bed, and not planted out the firft winter, their tap-roots will become fo ftrong and ftubborn, that they will be difficuit to manage in planting; and on the tap-roots being pro-perly difpofed depends the fuccefs of plant-ing in poor land.

IF there are any left that may not have been wanted, or that there is not time to plant, they muft not be left for another year, but muft be taken up, have their roots dreffed, and planted in the nurfery in the common method, to make other plantations in good land; for there fhould none be planted in poor ground, but from the feed-bed of that year's fowing,

THE planting in poor land is quite different from planting where there is plenty, or even a moderate quantity, of foil; for where there is fo little as two or three inches, if the roots are planted lower than that, there is nothing for them to ftrike into but gravel or rock, where it is impoffible for them to thrive.

THE holes being made as before directed, make them flat at top, and if they are two or three inches higher than the ground, fo much the better; for they will fink down level with the other ground in a little time, as their being higher is only from the earth's being ftirred.

TAKE up the trees from the feed-bed carefully, taking great care to break none of their roots, and take no more up than can be planted in a day; lay them in a flat bafket with fome mofs over their roots, and carry them into the field or common where they are to be planted.

OPEN

OPEN the hole five or six inches length-
ways, which may be done by only ftriking
the fpade into the middle of the hole after
it is flattened at top, which will be the
length of the tap-roots if they have throve
well. Lay in the root two or three
inches deep, as the ground will permit,
at its whole length, horizontally, and
then raife up the plant at the neck, in the
fame manner as in the planting of trees ;
faften the earth to the plant, keeping its
top upright ; after the plant is faftened,
loofen the top of the ground with the
fpade, this prevents it from cracking, and
is of great fervice ; for if the top of the
ground be left hard, two or three dry days
make it open as deep as the roots, and
dries them, and fo ruins the whole plantation.

IN light ground, where trees are raifed,
they will have but very few roots but the
tap-root, and that is the reafon that it is
made fo light to prevent fhort tufty roots :
they will be full of fmall fibres, none of
which

which fhould be cut off, and particular care fhould be taken to keep them from fun and wind, which would foon dry them, to the great detriment of the plant, for it is from them the tree begins to ftrike roots : as they are very fmall, they are very foon dried by either fun or wind. If it fhould happen by any accident they are, cut them off one inch diftant from the tap-root; but it will be more beneficial to the tree if there is no occafion for any of thefe amputations.

THE laying the roots at their whole length horizontally prevents the roots from ever attempting to run down into the gravel, which they would, if planted in any other method, and this is the reafon that plantations on gravelly and rocky grounds have hitherto made fo little progrefs.

THIS method of planting on poor land is entirely new, and as it is fo different from all kinds of planting hitherto practifed,

tifed, it may feem very wrong, as dreffing
and cutting the roots are the firft things
recommended by all that have given directions about planting. But let thofe
who have fuch ground to plant, only follow the directions given with accuracy, and
they will find them anfwer beyond their
expectation.

THE laying the roots horizontally prevents the trees from being raifed out of
the ground by froft; and as the roots will,
by this method, run juft below the furface,
they will have all the advantages it is
poffible for them to have in fuch fhallow
grounds; and this is the reafon that trees
raifed on poor foil are not fit to plant again
in the fame fort of foil, as their roots are
fhort, and feldom have a tap-root.

IT would be advifeable to plant fo that
the bend at the neck of the tree be always
againft the fun; for if the tap-root be very
ftrong, and there being no great depth of
earth to make it faft, it may throw the
top

top of the tree afide; if it does (which
feldom will happen if properly planted)
the fun will draw it ftraight; but if the
bend was to the fun, and happen not to
be upright, the tree would never be
ftraight. Were it not for that reafon, it
would be indifferent how the trees were
placed.

THERE fhould be two men employed
in planting; one to open the ground, and
the other to put in the tree, as it would be
a great lofs of time to lay down the fpade
fo often: they fhould have a little box
or bafket with a handle, into which they
fhould put a few trees at a time, with a
little mofs over their roots; this will be
very handy, as the man that plants muft
ufe both his hands in planting. This
work fhould never be performed when the
wind is high, nor when the air is frofty.

THERE is another method of planting
poor land that is either in fhort grafs or
fhort heath, which is very expeditious, and
also

alſo very advantageous, for it effectually
prevents the drought from penetrating in
ſummer, and the froſt from hurting the
roots in winter. Two men will plant five
or ſix hundred in, a day with eaſe. This ſort
of planting ſhould be done in winter when
the ground is full of wet ; the beſt ſeaſon
is after a deep ſnow is gone off with a
ſudden thaw; and at that time little work
can be done in any other grounds, as they
will be too wet to plant in.

THE trees are to be raiſed and managed
in the ſame manner in every reſpect as the
other trees for planting on poor lands, with
all the precautions concerning their roots.
A man with a ſpade makes a nick a little
longer than the root of the tree in the
graſs or heath, two or three inches deep,
as the ground will allow, laying in the
roots at the whole length horizontally,
preſſing the ground cloſe with the foot, and
with the ſpade nicks the ground croſs and
croſs all round the plant : this prevents
the nick from opening, which, without
that

that precaution, it would be very apt to do in fummer when the weather is very dry; it alfo prevent the wet from running off, which it does where there is a defcent, and the grafs is not nicked.

If the ground is very fhallow and not rocky, this is the beft method of planting; but if the ground is very ftony, it is not practicable, as the ftones would render it very troublefome to make the nicks, which interruption would often happen in the middle, fo as to prevent the roots from being laid horizontally or ftraight.

The trees for fuch plantations, that have not their roots drefled, fhould never be brought a great diftance, for it will be impoffible to prevent their fmall fibrous roots from drying, if they are kept long out of the ground; but if we cannot avoid bringing them from fome diftance, they muft be carefully taken up and packed in wet mofs: they will do very well for five or fix days; but if they lie much longer, their

their fmall roots will begin to mould, and muſt be cut off before they are planted, or their mouldineſs will deſtroy the large roots. If they come packed in moſs, they ſhould not be unpacked, but taken out as they are wanted, and planted as ſoon as poſſible.

It has been a common practice to ſend to the north for ſeedling-trees, where they are bought at a very cheap rate for planting out in poor land. In general the ſucceſs has been bad, and there are many good reaſons for its being ſo.

By chance there may be tolerably good ſucceſs, but it is a very bad practice ; for theſe trees are ſown as thick as they can grow, ſo that they have no free air, and by their cloſeneſs and their own warmth they are drawn up weak, and their ſtems are very tender, like a plant that is earthed up to blanch. This is eaſily to be diſcovered by obſerving the lower part of the ſtem, which is of a languid green for want of air, and the

the top quite of another colour where the air had a free paffage.

When trees thus raifed are taken from their clofe warm fituation, and expofed fingly to all weathers in poor cold ground, or in any ground, many of them will perifh, although ever fo carefully planted; and as all their roots muft be dreffed (as they will be either mouldy or very dry) they have none of the advantages of thofe feedlings raifed according to the directions before given. Thofe ftems are ftiff and ftrong, for as they were not allowed to grow nearer than three or four inches in the feed-bed, the air paffes freely amongft them, and their ftems are as hard as if they had been planted out three or four years, which is of great fervice when they are planted in the fields.

It is impoffible that trees raifed in the manner here directed can be fold fo cheap as thofe that grow fo thick as to be pulled up by handfuls from the feed-beds. But
6   it

it will be more to the purchaser's advantage to give ten fhillings a thoufand for trees thus raifed, than to give one fhilling for thofe that are fown and grow fo thick; there is more than that difference to the nurferyman, as twenty will grow in the thick feed-bed for one in thofe that are properly managed.

THERE are many thoufands of Scotch Firs fold in and near London, and many other places in the country, that are brought many miles, and are fome months out of the ground; thefe are fown in the feed-bed as thick as grafs. Befides the other difadvantages they have in common with other foreft-trees, as Oak, Beech, &c. they are very fubject to be rotted in their ftem even to the ground; and although they appear green at top, and the fmall roots they have are frefh, they decay and go off as foon as planted out and expofed to the air.

IF any fuch trees are purchafed, the beft

method

method to manage them would be to drefs their roots, and plant them in the nurfery at fix inches diftance, and in a year or two thofe that grow will be good plants ; but as many of them will decay, it would be troublefome and very expenfive to have them to go over for a year or two to make good the deficiencies, if they were at firft planted in a large common.

THIS is mentioned by way of caution, as fome gentlemen may think they have made a good purchafe by getting fo many choufand trees for fo fmall a fum, much lefs than the expence of raifing them would be ; but thofe that make fuch bargains will find to their great difappointment they have paid very dear, and had better have got good plants, if they had paid a much greater price.

WHAT has been treated of is only for the raifing of trees to plant out the firft year, from the feed-bed, into poor barren ground, and is different from the manage-

VOL. I.   G   ment

ment of raising trees to be transplanted from the seed-bed into the nursery, to be trained for some years to make plantations on different kinds of ground, where there is a sufficient depth of earth to make holes.

SUCH trees need not be thinned to so great a distance in the seed-bed as those for planting poor land, as they should be taken from the seed-bed and have their roots dressed, and planted in rows at a small distance, to stand a few years, and then to be removed again and dressed top and bottom, and planted at a greater distance, where they may remain until they are taken to plant in the fields to make plantations to grow for timber.

IT is a great advantage for all kinds of seedling-trees to be so thin in the seed-bed that the air may pass freely amongst them in the summer; they do not grow so tall, but there will be very few of them but will grow, and be much better when removed.

THERE are many gentlemen who are at the expence of trenching in all kinds of ground intended for planting. If it is a poor fhallow foil it is wrong, becaufe the gravel muft be thrown up or there can be no trench, and the little mould there is will be loft, and nothing left for the trees to ftrike roots in but gravel, mixed with a fmall quantity of earth, fo fmall that it will not be vifible in a dry fummer: they ftand but a poor chance to grow. If the trees were planted according to the directions before given, without trenching, they would thrive much better than it is poffible for them to do in the trenched gravel.

IF the ground be clay, and a fmall quantity of earth at top, it is of very bad confequence to trench; for to make even a very fhallow trench there muft be a good deal of clay thrown up. If the earth be mixed with the clay it will be loft; if laid in the bottom, it is turned down lower than trees fhould be planted in fuch ground, or the trench muft be fo fhallow as to be next to

110

no trench; therefore the trees in that cafe
will be planted in clean clay, and will
languifh many years before they reach the
good ground laid in the bottom in trenching, and often fo as never to recover, which
I have frequently feen. To plant in clay,
where the foil is very fhallow, the fame
method fhould be followed as planting in
poor gravel.

If the ground is good, trenching is attended with a great expence to no purpofe;
for if the trees are fmall or large, if the
holes are made fix inches wider than the
roots all round, for the young roots to ftrike
into, they will grow as well as if the whole
had been trenched, as the trees growing on
rocks where there is very little (and what
there is very hard) fully demonftrate; for
the roots twift and twine round the ftones,
and penetrate the earth between them,
which is extremely hard; as alfo by tree
roots going down into hard clay, which
fully fhews there is no occafion to trench
for timber-trees.

6   BESIDES

BESIDES the trenching there is another expence faved. Good land, when trenched, plowed, or dug, grows full of ftrong weeds, and muft be kept clean, or they will greatly impoverifh the ground ; and this will require to be performed for fome years.

I HAVE here deviated from what I pur-pofed, but as I have feen plantations made at a great expence by trenching, in all the different kinds of foil here mentioned, I have planted on all the fame forts of foils according to the directions here given, with better fuccefs and at a fmall expence.

G 3   CHAP.

CHAP. IV.

*On thick Planting, and the Management of
Woods that have been neglected.*

THICK planting has been recommended and practised many years;
but the method of dressing and thinning
is very seldom properly performed. The
prejudices against pruning are so great,
that we frequently see thickets of small
drawn-up trees of immense height, without a possibility of ever having good boles.

FOR the advantage of those gentlemen
that have such thickets within a possibility
of recovery, that is, from six years old to
twenty, the following directions will be of
use for making fine trees of what, in a few
years, would be past redemption,

IF a plantation has been planted five or
six years, and planted at three, four, five,

or

or fix feet diftance (for the diftance fhould be according to the goodnefs of the ground) and this neglected ever fince planting, we fhall find, if the trees have throve any thing like, they will all be mere thickets.

It will not be proper to remove any trees to plant in another place from fuch a thicket directly, for as they have been long fo very clofe, and have had no free air, they will be very unfit to remove, becaufe their ftems will be very tender.

First, let all the large branches be cut off clofe to the bole, and all the other parts of the tree be dreffed according to the directions for pruning: let this be done for two years, and then every other tree may be removed with fafety, and planted again. The pruning of the trees left in the old plantation muft be continued until the trees have got fufficient length of bole.

The trees that were removed, if they

were carried without earth, will want little this year; but if they had balls there will be little difference, and they muſt be treated accordingly, and they will ſtill make fine trees.

IF the plantation has been planted ten or twelve years, and has had no care taken of it all that time, if the ground be good they will be very tall and ſmall; and never will be fit for planting, for they would be ſo tall and ſlendeɪ, that they would never be able to ſtand the wind, nor ſupport their own heads.

WEAK trees taken from cloſe thickets ſeldom make good trees; for beſides their being tall, they have few ſide-branches, and ſhoot (if they grow) ſo faſt at top as to be moſtly crooked,

IF under-wood is of value, ſuch thickets are very fit to make a good wood of that kind, but it muſt be done with caution; for as the trees are tall and ſmall, and few ſide-fide-branches but juft at the top, they would be very liable to be much toffed and hurt by the wind if thinned too much at firft; befides, if they are expofed to the free air all at once, many of them will be in danger of perifhing, for they are very tender by being fo many years without a free air paffing amongft them.

If fuch thickets are defigned for timber-trees only, and were planted at three feet diftance at firft, let all the large branches of the trees be cut off the firft year, and none the next year.

Stub up every other tree, and mark the trees that are to ftand, and let them be pruned and dreffed according to the directions for pruning. The other trees that are to be taken up, fhould have the long branches cut off and ftand till next winter, and then ftubbed : by this method the trees will have the free air admitted amongft them gradually; and if they are afterwards

pruned

pruned regularly, in a few years they will thrive and be fine trees.

IF under-wood and timber is defigned, and that the under-wood is to remain, the timber-trees fhould be marked, and all the trees round them have all their fide-branches fnagged off a foot long, fo as to give air and harden the tree that is to ftand, which fhould have all the ftrong branches cut off clofe to the ftem.

NEXT winter all the trees round the timber-tree fhould be ftubbed, and it fhould be dreffed according to the directions for pruning. The winter following all the trees that are defigned for under-wood fhould be cut clean clofe by the ground; they will fhoot very faft and grow very thick. The timber-trees fhould not be left nearer than fifty feet to each other, for it is impoffible to have good under-wood if they are nearer.

IF

If the under-wood is only defigned to remain for a term of years, that is, until the trees grow up, the timber-trees may be left at ten feet, if the ground is very good; but if it is only middling land, eight will be fufficient,

A PLANTATION converted into timber-trees and under-wood that is to remain on good land, will be of more value than any other way it can be employed; for in countries where firing is fcarce it will every fourteen years yield a great profit to the proprietor, much more than good corn crops, if their whole value was calculated to the greateft nicety. And as the timber-trees are ftill growing, and will have very tall boles if they were managed as directed, they will be of very little detriment to the under-wood, and when they are of age to cut will be of great value.

If the plantation be feventeen or eighteen years old, never pruned nor dreffed, and planted at three or four feet diftance,

there

there muſt be great caution uſed to bring it into order.

THE trees will be very tall and ſlender, and have very few ſide-branches, and thoſe that are will be moſtly at the top and very long; their roots will be all intermixed, ſo that there will be no poſſibility of even ſtubbing any of them without hurting the roots of thoſe that are to remain.

THE only method that can be taken with trees that are thus tall and ſlender, is to cut out every other tree cloſe to the ground, and the long ſide-branches, ſome of which will be very thick, ſhould be cut off at a foot from the bole of the tree, that there may no blemiſhes be made in the bole.

THE heads of the trees ſhould alſo be thinned and lightened, for they will be all top-heavy, and be much hurt by the wind, which will have great power on them.

As

As they have had only every other tree cut off, if they were three feet diſtance at firſt planting, they will now be only ſix feet diſtance, which is too little if the land is very good ; but it will be beſt to let them remain ſo for two or three years, and by that time the trees that are to ſtand for timber will be hardened a little, and have got ſome little ſtrength.

BEFORE there are any trees cut, the beſt that are to ſtand ſhould be marked, and thoſe that are to be taken away may have their ſide-branches cut off at random ; but thoſe that are to remain for timber ſhould be dreſſed according to the directions for pruning. When thoſe that are to ſtand have got ſome ſtrength, the other trees ſhould be cut off cloſe to the ground.

IT will be many years before ſuch trees recover, and never will make ſuch good trees as if they had been dreſſed ſooner ; and this is all that can be done for plantations that have been ſo long neglected.

THE

THE trees that were cut off will fhoot ftrong, and make good under-wood for fome years; that is, till the trees get fome fide-branches, and grow to a fize to fmother it. But in fuch thickets, as the trees are fo tall, it would be very improper to thin them to fuch a diftance as to make a lafting crop of under-wood and timber-trees; for as they are very tall, and not ftrong in proportion to their height, the wind would fhake them fo much, if it did not break them, that they would be in great danger of being, what is termed by the wood-workers, *fhaken*, which greatly leffens the value of the tree.

THE under-wood will fhoot very faft, five or fix feet the firft year, and will be of great fervice to the ftanding timber-trees; in three years time it will be as high as the boles of the trees; but as their heads will always be above, it will not draw them, and they will grow very ftrong.

BEFORE

Before the under-wood is fit to cut,
the timber-trees will be a good deal proportioned in their bodies to their height,
and the winds will not hurt them when
the under-wood is cut.

As the under-wood will be very thick,
it will prevent any fide-fhoots growing on
the bole of the timber-trees, fo that after
the third year there will be no occafion to
prune them, nor in fummer to pull off any
of the fide-branches; for if there fhould
any fide-branches fprout on the bole after
that time, the thicknefs of the brufh-
wood, which will be very clofe, will fmother them, fo that they will decay next
year. There will then be no further
trouble.

All trees that have been neglected, and
have great fide-branches, although they
grow in avenues or fingle trees, may be
brought into order, fo as to increafe the
fize of their boles, but they will not be
very fightly, as all the large arms fhould

3   be

be cut off a foot from the body of the
tree; and it will be abfolutely neceffary to
pull off the young fhoots in fummer, that
fhoot from the places where the large
branches were cut off; for as they ftand
fo open, they will pufh many ftrong fhoots,
which will be more detrimental to the tree,
if not pulled off, than if it had never been
cut; for there are often to be feen trees
that have had large branches cut off, and
then neglected for four or five years, and
then cut again, with their whole bodies
one continued blemifh.

THIS is the error that the workers, in
wood fo juftly complain of, and it is in
general imputed to the pruning of timber-
trees; but it is not regular pruning, but
the neglect of performing it properly that
is the occafion of this difafter.

IT fhould be obferved, that none of the
directions here given will anfwer with any
of the turpentine kinds, as they never
fhoot after cutting; fo if there are any
thickets

thickets of them, the beſt way is to thin them gradually at two or three different times, and leave them as regular as poſſi- ble.

ALL the pine kinds will thrive and grow to fine timber at eight feet diſtance on the beſt of ground; and all the ſorts of firs at ten feet diſtance. The only way to have fine trees of thoſe kinds is to keep them thick, for they extend their ſide-branches to a great length when they have room to ſpread, and do not grow ſo tall and ſtraight in their boles as when they are confined.

PLANTATIONS of eighteen or twenty years old, that have been neglected and are very thick, may be brought to order by following the directions for that purpoſe, which will preſerve many good trees; but if it is in a country where firing is ſcarce, and of conſequence under-wood of great value, they would turn to more profit to cut them all down an inch below ground,

VOL. I.　　　H　　　for

for they will grow very faft and very thick.

In the beginning of July, fome ftools that have fine fhoots may be marked at forty or fifty feet diftance, of the kind the moft proper for the foil, or what the proprietor likes beft (if there are different kinds of trees in the plantation) and all the fhoots but one pulled off by the hand ; they will come off very eafily at that fea-fon, as the wood will be very thick.

In the winter, ftub a root or two the neareft to thofe that are to ftand for tim-ber, that they may have free air : they will make finer trees than if they had been treated in any other manner.

It will be neceffary to examine their bottoms next fummer, and pull off the fhoots, if any more have grown : they muft be pruned and dreffed as other trees ; and as they have free air and warmth they will grow amazingly, and in ten years time

time be much finer trees than they would
have been at the age if they had not been
cut down.

PLANTATIONS that have been planted
on tolerable or even very good land with
trees from the nurfery, of five or fix feet
high, and do not feem to thrive, whofe
bark looks reddifh, and pufh many fmall
fide-fhoots, and whofe leading fhoot often
decays, after they have been planted three
or four years, if they do not take to growing, cut them off an inch below the ground
in any of the winter months.

THEY will pufh many fhoots next fummer; in the beginning of July pull them
all off but one of the ftrongeft ; faften
the earth about the fhoot to prevent the
wind breaking it, which is the only danger it is liable to ; they will then grow
very freely, and foon be fine trees.

THIS method will anfwer very well for
all kinds of deciduous trees, the oak in
parti-

particular : they will fhoot a yard the
firft year, and be handfome ftraight trees ;
whereas before they did not fhoot two
inches, and even that killed in winter, becaufe it never came to maturity.

THIS muft never be attempted on trees
that are planted on poor land, becaufe all
trees on poor ground are feemingly hide-
bound for fome years after they are planted, and never make any progrefs until
they have been fome time at a ftand; if
they were to be cut down they would
make fhoots, but they would be very
weak ; the wood would not be ripened,
as it would be long before they fhoot in the
fpring, and they would be in danger of being killed in winter.

THE bottoms of all trees that are cut
down muft be carefully kept clean of all
fhoots ; they muft be pulled off by the
hand ; this muft not be neglected for two
or three years, for if the young fhoot that
is encouraged grows well, which there is

no

no doubt but it will, there will fhoots come from the bottom for fome time: thefe muft be pruned and dreffed according to the general rules for pruning and dreffing foreft-trees.

THERE are many kinds of wood very beneficial to be raifed where public works are carried on at or near the place, which occafions a great demand for wood.

WHERE there are coal-pits whofe roofs are bad, and require a great deal of wood for fupperts, it would be of great advantage in fuch places to allot twenty acres of good land to be fown with afh-feeds, which grow very faft.

AFTER they are come to a proper fize, if the ground was quite cleared as they were wanted, there would be an immenfe quantity of wood. They would be fit for fuch ufes in about twenty years after fowing; and if the field was begun at one end and cut regularly, before they were all

gone

gone over, that which was firſt cut would be fit for uſe, if the demand was not very great.

THE ſowing of aſh-ſeeds will be of great profit in ſuch places where much wood is wanted. The beſt method to have it come to maturity ſoon, is to ſow it in a field that has been corn a few years, the ſoil rich, and in good condition. Give it a winter fallow, and as ſoon as it is dry in ſpring plow it again, and let it lie until the beginning of April. The ſeeds being prepared, ſow them broad-caſt all over the field, and then a thin crop of oats ; harrow the whole.

THE corn muſt be cut high, and all the precautions that were given in the directions for ſowing amongſt corn obſerved. In the winter they ſhould be thinned where too thick, and in the ſummer, for a year or two, have the large weeds pulled from amongſt them, and they will require no more trouble.

As

As they grow thick there will be no occafion for pruning or dreffing ; only when any part of the field is cut in the winter, the next fummer all the fhoots, but two or three of the beft, fhould be pulled off from the ftools, and thofe left fhould be at as equal diftances as poffible. There will be no occafion to pull them any more, as the thicknefs of the wood will fmother what fprings after the firft fummer. This fhould be done every cutting.

THERE fhould never be any trees weeded out, for that, in a few years, would fpoil the whole field ; for as the wood will be very clofe, the young fhoots that fprouted from the tree which was cut would want free air, be drawn up weak, and grow crooked.

UNDER-WOODS of Beech and Hornbeam will be very profitable where much charcoal is wanted ; for if they are cut clean they will grow very faft from the ftools,

H 4 and

and be of a good fize. Although they be very thick, there is no occafion for doing any thing to them after they are cut ; only clear the ground of all rubbifh, and let them grow as nature directs, and they anfwer very well.

THERE are many compofitions recommended (and all of them would be a great expence if ufed for a large plantation) as fit to cover the places where large branches have been cut off; and that by applying any of them, they will prevent any blemifh in the tree, although the bough cut off be very large and cut clofe.

THIS I cannot agree to, as I have feen it often tried, and never found it anfwer. If any of them are laid on too hot it burns the bark all round the amputation, and makes the wound larger, and the bark rifes fo that it is long before it begins to cover. If it be laid on too cold it never joins well, but cracks and falls off. If

they

they are laid on juft of a right warmth, they will ftick until the bark grows over the wound, but there will be a dead place in the body of the tree, although the bark in time will cover the wound.

THERE can no large branch be cut clofe to the bole without making a blemifh. When the bark on the bole of any tree is bruifed, or rubbed againft by carts or any other accident, if the loofe bark is pared off immediately, and a compofition of equal quantities of clay and cow-dung be mixed fo thin as to be laid on with a brufh, as paint, all over the wounded place, and as foon as it is a little dry lay a plaifter of the fame compofition, made pretty ftiff, about half an inch thick all over, and dafh it with a little dry mould to prevent its cracking, there will be no blemifh.

BUT if the application is not made before the wound dries, where the tree was hurt there will be a blemifh, but the bark

will

will grow over better and fooner with this than any of thofe compofitions that are fo highly recommended, and at no expence.

THE laying the firft on fo thin and with a brufh is, that there may no part be miff-ed, for it penetrates into all the fides of the tree, clofe to the fides of the found bark, and it alfo prevents the ftiff plaifter from dropping off by the heat of the fun, which it would do if a thick plaifter was laid firft on the wounded place.

IF any favourite detached tree fhould have the misfortune to have a large arm or branch fplit, fo as to hang almoft by the bark of one fide, let all the fractured places be brufhed over with the thin ftuff, and then half an inch thick of the fame be laid all over,

RAISE the branch to its proper pofition, fo that the bark fits all round, and then fecured with props, fo as to prevent the wind fhaking it ; then apply a plaifter of

the

the fame, pretty ftiff, all round the branch
fix inches above and below the fplinter,
which fhould have a coarfe cloth lapped all
round to prevent its cracking, and faftened
with a fmall rope wrapped quite clofe. It
will unite and grow, and in a few years
be as found as ever.

THE compofition of cow-dung and clay
is better for all wounded places in trees
than any of the grafting waxes ufed for
that purpofe. Although this has no con-
nection with planting, it is often of great
fervice to trees that meet with accidents.

IF a branch is broken by the wind, and
flips off a good deal of the bark, lay on the
thin ftuff with the brufh firft, and then a
plaifter of the fame made a little ftiffer,
taking care to let it be an inch wider than
the wound, and be quite thin at the fides
to prevent its coming off by the heat of the
fun, and it will anfwer beyond expectation. There is no danger in laying it on,
and, if it is well faftened to the tree, it
will

will ftick until the wound is covered
with new bark. After the plaifter is laid
on, and a little dried, if it was brufhed all
over with fome of the thin that was firft
laid on, it would prevent its cracking and
fill up all the cavities, fo that no air could
get between the plaifter and the tree;
which if it does, the whole falls off, and
will be of no fervice.

THE dung and clay fhould be mixed
fome time before it is ufed, and worked
feveral times, that it may be well united;
if it was for three or four months it would
be the better, but there will be no occafion
to keep it moift all that time; for after it
is brought to proper order and laid in a
heap, it will require no further care until
a day or two before it is wanted.

C H A P.

CHAP VI.

On the Soils proper for the different Kinds of Forest-Trees.

THERE is a proper foil for all kinds of trees in which they thrive, and moft of the directions for planting ftop there. The variety of trees would be confined to a very fmall number if the fame kinds of trees did not thrive and grow in many different foils.

FOR the information and fatisfaction of thofe whofe foil is not the beft, and may be thought not to be fit for many kinds of trees, I have here added a lift of the common forts of foreft-trees, with an account of the different foils and fituations they will grow and thrive in.

THE obfervations were at firft taken from natural woods, and improved by fol-

3 lowing

lowing nature in many fuccefsful experi-
ments ; fo that if all the trees are planted
and fown in foils according to what is
here mentioned, they will thrive ; and if
managed after planting as directed for
pruning and dreffing, will make very fine
trees.

Oaks will thrive in clay if the acorn be
fown. They grow very flowly for fome
years, but they make fine trees, and are
the beft of wood. If planted from a nurfery where the land is good they feldom
thrive, but grow ftunted and crooked ;
but if planted from the feed-beds that are
made to raife trees for poor land, they will
grow very well, and after the firft year
fhoot freely.

In ftrong loam they may be planted at
fix or feven feet, from the nurfery, and
after the firft year will grow very well.

In light loam they may be planted from
the nurfery or other plantations when of a

large

large fize, and will grow freely even the firſt year.

In fat, rich earth they grow well at all ages, and make fine ſhoots ; if the acorn is ſown in ſuch ground they make the fineſt trees.

In black mooriſh land, where long heath grows, they grow faſter, and have finer ſhoots than in any other. The acorn ſown thrives well in ſandy loam ; but they muſt be planted young, or ſown ; for if they are large they grow hide-bound and never thrive. Acorns ſown round the bottoms of hills will thrive and grow fine trees, although the ground be very indifferent.

Poplars of all kinds will thrive in bogs and all moiſt grounds ; they will grow to a great fize in good land, but the wood does not deſerve ſuch quarters, unleſs they are planted by way of ornament for the variety of their leaves.

THE

THE American poplars are amazing faſt growers, and are beautiful trees, but the wood does not ſeem to be much better than the other ſorts.

ALL the kinds would anſwer well to be planted thick in moiſt grounds for coal-pit props, or where wood was much wanted for fencing of plantations and in-cloſures. The wood of late is much uſed in different kinds of huſbandry for its lightneſs; and ſome have made chamber-floors of it. The wood is of ſhort dura-tion, and is not of value to make large plantations in any other grounds but wet places, where other wood will not thrive, and there they will turn to great profit.

THE Alder wood is of ſo little value, and the tree of ſo little beauty, that it is not worth cultivating in any place but mere bogs, where little elſe will grow. They may be propagated in ſuch places by large truncheons drove or puſhed into the ground.

ASH

Ash is a tree of great ufe and value; in good land they grow very faſt. A gentleman may plant Aſh, and fee the fame trees cut and fold for three and four guineas each. It will grow in hard clay, but the wood is not very good; in ſtrong loam it will make fine trees; in light loam it grows amazingly faſt, and in fifty years after planting there will be trees of great value. It will grow very well in moiſt foils, but in hard dry ground the wood is not good, neither does it grow to a great fize.

The Aſh is not eſteemed, becaufe it is long before it is green, and lofes its leaves the fooneſt of all the foreſt kind. It is a wood much ufed in all forts of hufbandry work; fo that there fhould be a large plantation in good land, in a proper place that is not near nor much in view of the manfion-houfe, as they are very unfightly in fpring.

Beech will thrive in all kinds of foils

that are not wet; it will grow amongſt
hard rocks where there is hardly any ſoil;
it will grow in clay, but there they ſhould
be planted from the ſeed-bed, in the ſame
manner as trees planted on poor gravel.

If they are planted from the nurſery at
four years old they will do pretty well, but
if they are much older, there will not be
one in ten that will grow.

If they are planted in clay, or even in
ſtrong loam, they ſhould have only the
large ſide-branches taken off.

If they are trained in the nurſery, or
taken from other plantations, and have
been regularly dreſſed and pruned, they
may be planted in good light ſoil, when
they are large, with ſucceſs; but even then
they ſhould have all the ſmall ſide-branches
left on, for they do not thrive well when
they are much diveſted of their branches.
In the nurſery they ſhould not be much
pruned.

It

It is a tree of great use, and very profitable to make large plantations of. The favourite soil of the Beech is a dry light soil, of a foot or eighteen inches deep, with a gravel or stony bottom, and a high situation: in such ground it will grow very beautiful.

Birch will grow in any soil or situation, on the hardest rocks and in the softest bogs; those on the hard rocks never grow to very large trees, but the wood is pretty close; those in bogs and wet land grow to large trees, but the wood is spongy, and the trees will not live to a great age; those on middling land will make fine trees, and live to a good old age. The wood is tolerable, but is not of value to make large plantations of, unless where coal-pits or forges are near, to both which they are of great use.

They are also very beneficial to make common railing for fencing on wastes and commons that are planted or taken into

tillage,

tillage, becaufe they grow very ftraight and long, and are eafy to fplit when too thick. A plantation of them planted very thick on moift ground that is tolerably good would be advantageous, as they would grow very faft and ftraight, and would be foon fit for many ufes.

THE favourite foil of the Birch is a light black earth, of a foot deep, and a gravel or ftony bottom; in that they will grow tall and ftraight, with a remarkably white bark and fine heads.

THE Horfe-Chefnut will thrive in all foils where there is a moderate depth of earth, and a gravel or ftony bottom; there they will grow to a great magnitude. The leaves are handfome and the flowers are beautiful, but the wood is of no great value, fo it is only fit for pleafure.

A FEW clumps in the fkirts of plantations are very pretty. They fhould never be planted near walks nor houfes, as they
make

make a conftant litter from their firft budding until the leaves are all gone. It is a
tree not fit to plant for profit.

THE common rough-leafed Elm will
thrive in moft foils but hard clay and
ftanding water; yet it will thrive by river
fides, where the foil is light and fandy (as
it generally is) although the roots are three
parts wafhed bare by the water; but the
wet is only temporary and goes off foon.
It is an excellent foreft-tree, and is of
great utility both in the hufbandry way and
cabinet work; and it has this advantage
over all the other kinds of Elms, that it
will thrive very well in indifferent ground.
Its favourite foil is a light black earth, of a
loofe nature and a hard bottom; in fuch a
foil the roots will run a great way, and the
trees will grow to a great fize, and be very
ftraight, if planted thick.

THEY are raifed from feed and from
layers; many prefer the feed, but if the
layers are rightly managed there is little

difference,

difference, if they are moved once or twice in the nurfery: they may be planted out from thence, or from other plantations that have been properly managed, after they are large, with good fuccefs. It is an excellent tree to make large plantations of, and will turn to great profit, as it will thrive fo well on very indifferent ground.

THE Dutch Elm is a tree that grows very faft when young, if the foil be good; and as it fhoots immenfe quantities of fide-branches, they require a good deal of labour to keep them in order. It likes a good depth of foil, a good rich loam; there it will thrive and grow very faft for twenty-five or thirty years, but after that time it generally is at a ftand, is ftunted, and makes no figure; the bark grows rough, and the wood is of little value, fo large plantations of them would not be very advantageous, efpecially as they require good land, which may be better employed.

THE

THE Englifh Elm is the moft beautiful of all the foreft kinds, and grows to an amazing great fize. It will grow in a ftrong loam where there is a good depth of foil above the clay, which is generally the bottom where the top foil is a ftrong loam.

As all thefe trees are raifed from layers, they fhould be trained in the nurfery for fome time before they are planted out: if they are properly managed, the roots will be fo trained as to run horizontally, and if they are fo ordered, they will grow to very ftrong trees in a ftiff loam; for after the roots are flat they will run along juft below the furface, and never attempt going into the clay, which is a great enemy to them, for they ftunt as foon as the roots touch it.

THEY will thrive very well on a light black earth, and on light loam that is of a moift nature they thrive the beft of all: fuch loam has always a fandy bottom; and

I 4   although

although the Elms are trained so that their roots are quite flat, and the large ones as they extend run just below the surface, there will be many small roots a great depth in the sand. They will thrive surprizingly in soft sand, but then they should be planted young and pretty deep.

THEIR leaves are not of that dark green when in such sandy land; they lose them sooner in autumn and are out earlier in spring. It should be observed, that their situation be on the lowest ground, that the moisture may fall to them; for although they do not love to stand in wet, yet they like a good deal of moisture, and they always thrive best in that situation.

IT is a tree very fit to make large plantations of, as the wood is very valuable, and will turn to great profit, if the soil be proper and of a good depth; but on poor land and shallow soils they thrive the worst of any of the forest kind.

THE

THE Witch Elm is generally miſtaken
for the common rough-leafed Elm, which
grows wild in moſt woods, and is fine
wood; whereas the Witch Elm is a ſoft
ſpongy wood, not much better than a Sal-
low, and grows almoſt as faſt : it grows
very tall, and ſmall in the bole to its
height. It will thrive in an indifferent ſoil
if it is pretty deep ; it grows beſt in a
good light loam. It is only fit for hedges,
to cover arbours, or (planted very thick)
to cover old walls in ſummer, or any other
diſagreeable objeᴄt, for all which it is very
fit, as its branches and leaves are very thick.
It is not worth cultivating for any other
uſe.

THERE are ſome other varieties of varie-
gated Elms that are for beauty and ornament in pleaſure-grounds, but of no great
value as foreſt-trees, ſo are of no conſequence at preſent. They are propagated
from layers, and budding on the common
rough-leafed Elm. They require a good
deep ſoil, and not too rich, or they loſe
much of their beauty.

THE Lime tree is a very beautiful growing tree, and is very fit for fhady walks
and clumps for ornament, but, like the
Horfe-Chefnut, makes a conftant litter,
which is the reafon it is not now in much
efteem. Its wood is foft, and of no great
value, but for carvers and fuch trifling
works, fo is not a fit tree for large plantations. It will thrive in foft fand, and any
light foil that is dry : its favourite foil is
a fandy loam, in which it will grow to a
great fize, and if it has room to fpread
will be very beautiful.

THE Sycamore is a fine growing tree,
and the wood is of great ufe for turners,
and many other things in the furniture and
hufbandry way ; it is alfo a very ornamental tree in plantations, and in clumps
where the ground is fit. It thrives beft in
a fandy loam ; it grows very well in all
light dry foils that are of a moderate depth.
It is worth propagating for profit, where
the foil anfwers.

THE

THE common Maple will thrive in all poor dry grounds, even where there is but little earth: it is a wood of no great value, and feldom grows to a great fize, and is not fit to make plantations, neither for profit nor beauty; it is very good for to convert into under-wood, as it fhoots very ftrong and very quick from the ftools after the firft cutting.

MAPLE (the Norway) is a good tree, the leaves are handfome, and make a fine fhade: the wood is much of the fame nature with the Sycamore, and will thrive in all the fame forts of foils, It is a very good tree to plant for ufe.

THE Mountain Afh is a very handfome tree; the leaves, flowers, and fruit are all very pretty: the tree grows to a good fize, but the wood is of no great value. The leaves and flowers make a very beautiful appearance in fummer, and in the autumn, when the fruit is ripe, they are of a fine red colour. If three or four are

6 planted

planted in little clumps in different parts of the plantation they would be very ornamental; it is not a tree to plant for profit. It will grow in very poor land, and thrives in moſt ſoils, even amongſt rocks and in ſtiff clay.

LABURNUM (the Scotch) is one of the moſt beautiful trees of the foreſt kind; its bark is remarkably green, with a fine ſhining leaf, and when in flower is all over a bright yellow. It grows to a great tree if planted in good, light, rich earth; it will thrive in all light ſoils that are moderately deep.

IT is a very valuable wood, and will yield great profit to make large plantations of, if near a great town or convenient for water-carriage, as it is too beautiful a wood to uſe for any work but the cabinet buſineſs.

FOR clumps in pleaſure grounds, Laburnums and Birches would be very beautiful,

as

as the bark of the one is white, and the other a dark green : they are both much of the fame growth as to height, and both their leaves are very fine.

The Spanifh Chefnut will thrive in all good foils that are not wet ; it will grow in fandy loam, and in very poor dry ground; but then it muft be planted young, that is, from the feed-beds, as di-rected for planting poor land : it may alfo in fuch foil be propagated from the nut. It makes a fine fhade, grows to a very large tree, and is a good durable wood: it is ufeful for all kinds of country bufinefs, and is valuable to make large plantations.

It may be objected that fowing the nuts or tree-feeds on poor land is contrary to the directions given for planting poor land, and the reafons that are given for its being done in that manner are, that all feeds *fown* go down with a tap-root, which they certainly do if the ground is good, or if even it is a hard clay ; but in gravel or ftony ground the tap-root cannot get far

3   before

before it meets with fome obftruction and
is turned afide, after which it fpreads and
runs amongft the ftones and along the top
of the gravel, as may be feen in natural
woods.

TREES that have been felf-fown in ftony
ground, their roots run many of them on
the top of the ground; or if they are
covered when young, when they grow
large they rife above the furface, fo that
moft of the roots are feen above ground,
and they run the fame as if they had been
planted and laid horizontally; fo that
planting and fowing in bare ground is no
ways contradictory.

THE Walnut: its proper foil to plant
in is a fine, rich, light, black earth of two
feet deep (if it can be had) with a gravel
or ftony bottom; in fuch foil the trees will
grow to be very large and bear great quantities of fruit, and very good. It will thrive
on a ftrong loam, if there be a good depth
of it, but it will not bear much fruit,
and what there is will not be good. It
will

will grow very well in a fandy loam, and bear great crops of fruit, which will be fmall, but very good.

It will grow and thrive very well as a timber-tree on all deep foils, although inclining to clay, but will bear very little fruit, and thofe will not be very good. It is a wood of great value, and in proper foils no tree yields more profit. It is a too beautiful wood to ufe in any bufinefs but furniture and ornamental works, fo that if they are not near a market it leffens their value, which fhould always be confidered where timber-trees are planted for profit.

CHERRY (the common Black or Wild) is an exceeding good foreft as well as fruit-tree, and in places where diftilleries are near, their fruit yields great profit. It grows to a great fize in all kinds of good foils, and will bear fine fruit equal in flavour to the fineft cherries. It will thrive in very indifferent ground, provided it has a dry

bottom

bottom (for wet is a great enemy to it) but the fruit will be ſmall. In ſuch ſoils it ſhould be planted from the ſeed-bed, as directed for poor ſoils.

Iᴛ will grow in all ſoft moiſt ſands, and in ſtrong loam it will grow to a great tree, but the fruit will be hard, dry, and bitter. It is a tree that will thrive in almoſt any ſoil but where there is ſtanding water. It is very beautiful when in bloſſom; ſome clumps of them properly placed in large plantations would add great dignity to the whole; and as the wood is very valuable, they deſerve to be encouraged in all plantations.

Pʟᴀᴛᴀɴᴜs or Plane-tree (occidental and oriental) are two ſpecies differing only in the ſhape of their leaves; and as botanick diſtinctions are no part of the plan of this work, but plain and eaſy directions how to manage and bring to perfection all the kinds of trees mentioned, I ſhall claſs them together, as their wood and culture

are

are much the fame. They are very beautiful, have fine leaves, and grow to a great magnitude. There have not as yet been any large plantations made of them, but there are many fine trees growing in feveral parts of England.

THE wood is much the fame as that of the Sycamore, and they will thrive in much the fame kinds of grounds ; but to have them grow to perfection they fhould have a good, rich, deep loam, not too ftiff; in fuch foil they will grow to an enormous fize, and when fuch a foil happens in a plantation, fome clumps of them would add greatly to its beauty.

THE Hornbeam is much of the nature of the Beech, and will thrive in all the fame forts of foils. I can affign no other reafon for its being generally neglected in plantations, than that the Beech is a much handfomer tree, and grows much ftraighter in the fame kinds of ground and fituations. The Hornbeam, if planted thick, will grow

very ſtraight and tall, and make a fine tree.
There is no wood fitter to cut down for
under-wood; it grows very quick and very
thick.

THE Crab is a tree that grows to no great
ſize, and large plantations of them would
not be prudent. They are a fine, hard,
durable wood, and are uſeful, eſpecially
for many things belonging to Mills. It
has the fineſt bloſſom of all the flowering
kinds, and a few of them properly diſ-
poſed in a large plantation would be very
beautiful.

THEY thrive beſt in a good rich loam,
and there will grow to a pretty good ſize;
they will alſo grow in very indifferent
ſoils that are dry. If they are planted
thick they will grow tall and have fine
ſtraight ſtems; but their greateſt beauty is
their fine ſpreading heads, ſo if they are
planted for beauty they ſhould have room
to ſpread.

THE

THE Larch tree is a very beautiful plant, and makes ample amends for its not being an evergreen, by its fine appearance in the fpring. Moſt plantations of them that have been planted are in low grounds, good land, and warm ſituations, which is very wrong; for in ſuch places they grow too faſt, like a plant that is drawn, under glaſs, for want of air. Such places are not their proper ſituation: as they were of foreign growth, it was imagined they were tender, which they are not.

THEY are very hardy, will reſiſt the greateſt cold, and will thrive exceeding well in high ſituations, and in very poor land, and there will grow to fine large trees and good wood, which they never will in rich ſoils; for they grow ſo faſt that they cannot ſupport themſelves, and ſo are always crooked. If deſigned for wood, they ſhould be planted in clumps pretty thick, for they ſpread prodigiouſly when they have room or are detached, or planted promiſcuouſly amongſt other trees, which

K 2 they

they overgrow in a little time; this has been the common practice in planting them.

THERE is no tree they can be mixed with to succeed but the Mountain Pine; they are the moſt agreeable companions, for they are inhabitants of cold high mountains, and in ſuch ſituations will grow very equally.

THE Larch will thrive in any ſoil or ſituation; it will thrive very well in a ſtrong clay, and I have ſeen them growing very freely where one third of their roots were in the water. This was in an iſland, the ſurface of which was not a foot above the level of the water, and of conſequence moſt of the roots were in wet ground; ſo that they may be ſaid to grow in any place where there is mould for them to ſtrike root. If they are planted in clay they ſhould be planted in the ſame manner as trees on poor gravel. They ſhould never be removed when they are large, for they

thrive

thrive badly for fome years if they grow, which is very hazardous, although they are taken up with large balls.

I HAVE been told that the wood is very good (but there are no trees come to proper age in England to judge of its goodnefs) and that it hardens in water; for if it is always wet, it will laft for hundreds of years; which, if true, it would be worth propagating, if only for the ufe of waterworks.

THE Common Spruce, the Norway Spruce, the White and Black Spruce, are all different fpecies of the fame tree, and the fame management and culture will anfwer to bring them to perfection. They are all fine trees and grow to be very large, and when our plantations of them come to maturity, we fhall have as good wood as any that is brought from Norway or any other part of the globe; and as there are great quantities of ground fit for their growth that is of little value, they fhould

K 3

be

be cultivated with great care and induſtry.
They will thrive in poor gravel and ſtony
mountains; on ſuch ground they ſhould
be planted from the ſeed-bed, two years
old.

In marſhy, mooriſh ground they will
come to great perfection, as in ſuch places
the ground is wet and the ſoil deep. It
would be beſt to plant them when a foot
or two high; that is, plants that have been
removed from the ſeed-bed into the nurſery,
and have ſtood two years. They will even
grow in wet moſſes, but there the trees
ſhould be four or five feet high, and the
beſt for that purpoſe are trees that have
been ſome time planted, and growing on
hard bare dry grounds.

They may be planted at firſt at ſix feet
diſtance, and ſhould not be moved until
they meet, and then every other tree ſhould
be taken up and planted in other places;
thoſe taken up in the hard dry ground will
be very fit to plant in the wet moſſy
grounds,

grounds, and the proper feafon for planting them is the laft week in March or the
firft week in April.

THE trees taken from the hard dry
ground, their roots will be fhort, tufty,
and fmall, none of which fhould be cut,
but planted again as foon as poffible, and
it would be right to plant them a foot
higher than the level of the ground, if it is
very wet; they will ftrike root and grow
immediately, and will have good hold before winter, and then the wet will not
hurt them. As the trees are large that are
planted in the marfhy and moffy ground,
they may be planted at twelve feet diftance
at firft, fo they will require no further
trouble.

SILVER Fir is a beautiful tree and grows
to a great magnitude, but muft have better
ground than the Spruce. If it is planted
in a light black foil, after the third year
from the feed, it will grow very faft; it
will thrive in light loam, and will grow

K 4   in

in fandy loam, but not to a great fize: it makes a handfome detached tree, and will fpread its branches a great way, and will be very ornamental. If a plantation for timber, twelve feet will be fufficient, and they fhould be planted in large clumps.

I have feen Silver Firs of fixty years old that were almoft a hundred feet high, and thick in proportion. The foil was light and fine, not much more than eighteen inches deep, and not very rich.

The Pinafter is not much different in its appearance, at a diftance, from the Scotch Fir, and while young they feem to be much the fame; but as there is no plantation of them old enough to try the goodnefs of the wood, it cannot be determined if it is any better than the wood of the Scotch Fir; it is equally hardy, and will thrive in the fame foils and fituations, and may be planted round the fkirts of plantations to fhelter other trees.

The

THE Swamp Pine has only this advantage over the others, that it is fuppofed to grow in wetter places than them. As to its wood and beauty, it is, to view, not much different from the Pinafter, is as hardy a tree, and will thrive on hard dry ground as well as in bogs. The Scotch Fir will grow in very wet ground.

THE Mountain Pine (commonly called the Scotch Fir) is what we are beft acquainted with, and is known to thrive in poor gravel, rocky mountains, and in all kinds of foils and fituations; but in low fituations and good land the wood is not fo good as when the foil is but indifferent and the fituation high, which makes it valuable.

I KNEW two plantations of Scotch Firs that were planted at the fame time; fixty years after they were planted they were cut. The one was on a rifing ground, which was fhallow and very poor; the other was on a low ground, which was ftrong, and all the winter covered with water: that

on

on the low ground the trees were in general
four inches more in diameter than those on
the rising poor land, but the wood was not
so good, it was white and spongy ; the
other was firm and of a good red colour,
and of much greater value.

The Weymouth Pine is the beautifullest
of all the Pine kinds, but is tenderer than
any of them; by the growing of the wood
it seems to be finer than any of the others,
but as yet there have none been cut of a
sufficient age in England to know its goodness ; neither have there been many large
plantations planted of them, for what has
been planted are mostly for pleasure, and
most of them in good ground and warm
situations, where they have throve very
well : they will also grow in indifferent
soil, but should be sheltered from the north
and west winds.

C H A P.

CHAP. VII.

On American Foreſt-Trees.

MOST of the common kinds of foreſt-trees in England have been treated of, and the manner of propagating and bringing them to perfection; but there are numbers of American tree-feeds brought over every year and ſown in England, many of which are fine plants and pro-miſing.

IN the year 1772 I had a collection of tree-feeds from America, amongſt which were many ſorts of Oaks, which were ſown in very indifferent ſoil, but a warm ſituation; moſt of them grew very well, and made ſhoots much ſuperior to what I expected in ſuch ground.

THE woods in America are, in general,

on

on very rich, fine, deep ground; and by a
gentleman who was very curious in planting and obferving the progrefs of the
growth of wood, and who had refided
many years in America, I was affured
that trees of all kinds, in that country,
grew as much in one year as they did in
England in three. He alfo affirmed, that
if a wood was ftubbed, and brought into
corn, and kept fo for fome time, and then
laid into grafs, in a few years it would be
all a wood again.

It is well known that it is not fo in
England; for if a wood is deftroyed and
taken into corn, and kept for that ufe for
fome years, when laid into grafs, if it continues never fo long in grafs, there grow
no timber-trees; thorns, broom, and furze
will grow, and cover a field in a few years,
if not prevented, where fuch fhrubs are
near, and even fometimes where there is
no fuch thing, but that is eafily accounted
for, as the wind and birds may and do
carry the feeds; but timber-trees never
grow

grow there, unlefs a wood is near that has trees bearing feeds, which may fcatter their feeds there as well as in any other field.

I AM far from being againft the introducing and propagating of foreign trees, as many of them are great beauties; but I would not have planters be too fanguine, and plant large plantations of foreign trees on the recommendation of foreigners, for all or moft nations are partial to their own country, and recommend the produce as far fuperior to what they fee in other places, without confidering what it is that makes the difference.

IT muft be many years before the goodnefs of the wood can be determined by young trees raifed from the feed and planted in England. The beft method would be to procure fome wood that has come to maturity in the country the feed comes from, and of the fame kind, and it would be neceffary to know what kind of foil

and

and situation the wood grew in, for the soil alters the wood very much.

IF the wood is good and fit for any particular use, then it would be worth the while to plant the foreign and some English trees of or as near the same sort as possible, in the same soil and situation, in clumps, the trees planted alternately; this would be a fair trial as to growth: then it would be right to consider whether this foreign wood is better than the English that grows on the same soil; if it is, and grows faster, it is a great acquisition, and fit to plant for profit.

SOME years since I raised from seed the American and the common Spruce Fir; the American grew taller the three years they were in the nursery, but not so strong as the common: they were then planted into little clumps in a rich light loam. The first year neither of them made much progress, although they were taken up with good balls of earth, and were not a quarter of an hour out of the ground.

THE fecond year the American made a
fhoot of two feet long; the common a
fhoot of fifteen inches, but near twice as
thick.   The third year the American fhoot
was much the fame; the common twenty-
two inches and very ftrong.   The fourth
year the American kept ftill to two feet;
the common gained an inch.   They both
grew much the fame for many years; the
American was the talleft, but the common
was much the ftrongeft plant.

THERE were the fame year, and near
the fame time, fome planted on a dry,
poor, fandy foil.   They both made very
poor fhoots for feveral years, but the common was the beft, and of a much darker
green.

I HAVE taken notice of this difference,
becaufe moft that have wrote on planting,
of late years, greatly recommend the American kind of trees, as being of a much
quicker growth than the fame kinds that
are Englifh; that is, that an American
Oak,

Oak, Afh, Beech, &c. will grow much faſter than an Engliſh Oak, Aſh, or Beech.

I HAVE at this time Oaks of different kinds, Aſh, Beech, and Birch, from American ſeeds, and the ſame ſorts that are Engliſh, growing in the ſame nurſery; there are ſome of both ſorts that make long ſhoots and ſtrong; and in many of them there is no material difference. I believe it is more from the richneſs of the ſoil than from the trees, that cauſes the great difference in the growth of wood in England and America.

THERE are many kinds of American oaks that have been raiſed from ſeed ſome years ſince in England, and are planted out, and are fine thriving trees, and grow much faſter than the generality of the Engliſh; but as they were new trees, and greatly recommended for their beauty and quick growth, they were ſown in the beſt

beſt ſoils, and planted with all care in the warmeſt ſituations.

BUT if they had been planted in ground of leſs value, perhaps they would not have grown faſter than the common Engliſh oak ; and if they did, it would be very material to know if the wood was equally good, for the quantity will not make up for the quality, if it is of an inferior nature.

EVERY one knows that all the kinds of foreſt-trees we have in England, that are of quick growth, their wood is of very little value, as poplars and willows of all kinds ; but there may be ſeminal varieties in ſeedlings of all kinds of trees, as the quality of the wood may be equally good and yet grow much faſter ; for in large plantations of all kinds of trees there will be many that will ſhoot as much in one year as the generality of the plantation do in two or three, where there is not the leaſt viſible difference in the ſoil.

VOL. I.   L   WHE-

WHETHER this is owing to a seminal variety, or to some superior quality in the ground where such plants are planted, I do not pretend to determine ; but if, upon the nicest observation, there cannot be discovered the least variation from the others that grow so slowly, I think it would be surprizing if it proceeded from a change in the seed, but rather from some hidden quality in the soil.

THOSE that are curious in trees, when they observe a seedling-tree that is remarkably different, and has some good qualities which they would wish to preserve, the best method is to convert it into a stool, and raise plants from layers (for all forest-trees will grow from layers) which is a more certain method than grafting ; for all trees partake something of the stock they are grafted on, as can be easily proved from the dwarfishness of the apples grafted on paradise stocks, and pears on quinces, with many others.

THOSE

THOSE that are accurate in botany can diftinguifh many fpecies from feedlings, fown at the fame time, of any kinds of trees or plants ; and it is owing to their ingenuity and difcoveries that makes us poffeffed of fo great a variety of beautiful plants of all kinds.

THE only difficulty in feedling-trees is to know which is the beft wood ; as in all kinds of feeds fome improve, fome degenerate, and many keep to their kind. But when there are many vifible differences in plants from the kind that was fown, there is no doubt but the very nature is altered ; and although they may feem to be improved by being more vigorous than the others, perhaps the wood may be much inferior.

IN flowers and other inferior plants, their goodnefs is foon difcovered, and in general (efpecially in flowers) the ftrongeft and freeft growing are the worft; but in trees it will be a long time before it can be

known,

known, not before the trees are fit for
cutting; for there are many kinds of wood
which is good for nothing until it comes
to its proper age, which is then very good;
and many kinds that are pretty good when
young, but when old are good for little.

For which reasons it will not be prudent to plant large plantations for profit
with trees, let them be never so well recommended, until the goodness of the
wood is well known.

I have been informed by several gentlemen that have been many years in America, that in the large and fine growing
woods there the soil in general is a fine
light loam of a great depth, such as is in
our valleys by river sides, and are perfectly
dry.

If a large plantation was made on such
ground in England (for if there are only
a few trees planted they will never make
fine timber) and if the water does not lay

on

on it in the winter, which is often the cafe, they would grow very faft and be fine trees. Some years fince I made a plantation on fuch a foil, which was not liable to be wet in winter.

THE plantation was moftly the Englifh oak; they in general made fhoots two feet long the fecond year after planting; the fhoots were proportionably ftrong. They continued to grow very quick, fo that in a few years they were fuch large and fine trees, that none who faw them would believe but they were much older.

THERE are now growing at Lord Downe's at Cowick, in Yorkfhire, what are there called the large American oaks; they are growing in fmall clumps; they are about thirty years old; the moft of them are fifty feet high and upwards, two feet diameter at bottom, fine ftraight ftems, clear bark, and grow freely.

THE feeds were fown where they are

now growing; they were kept clean and dreſſed carefully for ſome years. There are no Engliſh oaks growing near them; if there had, I am of opinion they would have been as large as the Americans, for there are ſome Engliſh elms growing near them which are much larger; which makes me believe it is more the ſoil than the kinds that makes the quick growth of trees. His lordſhip's park is a rich, light, deep loam, and a warm ſituation.

CHAP

C H A P.   VIII.

On the Management of grown Wood.

ALTHOUGH the intention of this publication is only for the planting of poor lands, and converting them into profitable forefts, as fuch ground is of very little ufe and of no value, it cannot be fuppofed that trees on fuch ground will ever grow to the perfection and fize of trees on rich foils ; fo that every gentleman of landed property fhould allot a few acres of good land for planting for the good of pofterity, and the advantage of his country.

IT will not turn to the planter's immediate profit ; but there are few of that parfimonious difpofition who will do nothing for futurity, who are really gentlemen ; and all thofe that are lovers of trees, as

L 4   moft

moſt gentlemen are, their pleaſure in ſee-
ing them grow will be very great, and
give them more joy than any other amuſement. *Theſe are trees of my own planting*
are words I have often heard repeated with
great content and gladneſs.

ALTHOUGH I have treated a good deal
already on the thinning of plantations on
good land, I muſt again repeat a caution.
The methods of planting good land is
ſkilfully and plainly taught by many able
authors, ſo as to want no inſtructions but
ſome cautions.

PLANT thick and prune carefully according to the directions given for pruning, and juſt when the ſide-branches begin to meet remove every other tree, for
if they ſtand until they are become a
thicket, it makes them tender for want of
free air, and when expoſed are in danger
of being loſt. The removed trees ſhould
be planted in large clumps, or in a new
plantation ; for trees from ſuch places
ſhould

fhould never be planted as detached trees, nor in fingle rows.

THE trees that are taken up muft not by any means be divefted of all their fide-branches, as is often practifed; for if they have been well managed, they will be ftiff and ftrong, and very able to fupport themfelves with the fide-branches on ; they keep the trees warm, and encourage their growth greatly; whereas, when taken clean off, the top or leading fhoot grows very long even the firft year after planting, and always fo weak, that it is not ftrong enough to fupport itfelf, fo grows crooked ; but where many or all of the fide-branches are left on, it grows not fo much in length, but quite ftrong and upright.

IT is owing to the ftripping of young ones that there are fo many crooked trees in plantations that are planted from nurfe-ries with trees eight or ten feet high.

I HAVE planted trees twenty feet high

with fo many of the fide-branches on, that
by the middle of fummer it would have
been very difficult to know at a little diftance, but that the trees had been growing in that place feveral years.

IF trees that are very tall and have
ftiff boles are to be removed, if they are
ftripped (which is very often the cafe) of
all their fide-branches, they grow to a
great bufh at top, and are in danger of being broke ; but if there had been many
large branches left pretty thick on the
bole, the tree would bear the wind much
better, as it would then have all bent
equally, been in no danger of being broke,
and made a finer tree.

THE difficulty of planting trees that
have been a long time in a thicket, is,
that they have very few fide-branches, and
thofe that they have are very long. They
fhould be fnortened to two feet long, and
the top, which is generally very bufhy,
fhould be thinned by cutting out fome of
the

the largeft clofe to the bole, and fhorten-
ing the others gradually, fo as to draw the
tree to a leading fhoot.

BUT it is bad planting them; for as
they have been fo long deprived of free
air, and been very warm, they are in
great danger of being loft, fo that it is
only in cafe of neceffity they ever fhould
be planted ; and when they are, the only
place is in a wood, where there is a large
vacancy by a tree's being dead, or cut out
for fome particular ufe. There they may
fucceed, but hardly any where elfe.

IN fuch places they are of great fervice,
and it is difficult to find trees that are tall
enough but from fuch thickets, and there
they are in no danger of being fhaken by
the wind, as they are protected by the trees
all around.

IF the vacancy is large, that is, forty
or fifty feet diameter, which is often the
cafe where a large tree has been cut out, it
would

would be right to plant four or five in the middle pretty clofe ; for it would be to no purpofe to plant within twenty feet of the fide, unlefs the tree that is planted is taller than the trees that are growing round the vacancy.

It is a common practice in natural woods to cut out trees for particular ufes : this is very wrong ; for it would be of more advantage to the owner to purchafe what wood is wanted, even at a very dear rate, than mangle his wood ; for if the wood was not well managed when young, it is probable the cutting out of a large tree may make a gap of twenty or thirty feet.

No young tree can be got up even in that great fpace, for the trees that are round will fpread their fide-branches, which were kept up by the tree that was cut, and in a few years will meet ; fo that if even a tree from fome other place was to be planted in the opening, unlefs it was

3   higher

higher than the trees round it, it would
foon be over-hung and deftroyed ; fo that
there is not the leaft probability of any
fhoots that come from the ftool of the tree
that was cut off ever getting up fo as to
come to perfection, even although all the
fide-branches of the ftanding trees were
cut, which would be a great detriment to
them, more than the value of the planted
tree, even if it fhould thrive very well. It
is owing to this bad practice of cutting
out trees, that there are fo many crooked
trees in natural woods.

THERE is another great objection againft
the cutting of trees in grown woods.
There are generally cattle allowed to
feed in them, and if the owner caufes
fences to be made round where the tree
was cut, to fave the young fhoots (which
I have feen) they are little regarded after
the firft making ; and indeed if they were
it would be to little purpofe, for if the
young fhoots were never fo well preferved, before they could come to any height
the

the place is quite clofed at top, and they grow crooked, unfightly bufhes.

IF the wood is come to its full growth, and is to ftand fome years, it is a great lofs ; for when the whole of the wood is felled, the fhoots from the tree that was cut before will be fo bent by the over-hanging of the trees, that they will be fo crooked as to fpoil many of the young trees round them, if they are not cut down, and if they are, it is wafting the ftool to no purpofe.

IF the wood is fit to cut, and is left be-caufe it is beautiful, and the owner has no mind to part with it, the vacancy muft either remain empty or be filled with a crooked bufh, which are fufficient reafons to put an end to that bad practice.

IF a little wood is wanted, let fome corner that is moft out of fight be cut down and afterwards planted, or the ftools managed as before directed, and then there

will

will be woods both beautiful and profitable. No gentleman should allow their woods to be cut and mangled, as most of them are in the north.

It is a custom when oak woods are cut to leave many small trees, which is a bad practice ; for if the wood was thick and good, as it ought to be (and would have been, if well managed when young) these small trees will have long slender boles, and are very liable to be broke by the wind; if they should escape being broke, they are often so twisted and shaken, that they never make good wood. If they should meet with none of these misfortunes they will grow to great bushy heads, and over-hang many of the young trees.

When a wood is cut, if there are any stiff young trees, and four or five of them can be left in a clump, it will answer very well, and they will grow fine tall trees ; but if they are left detached they will grow no taller. If they meet with no mis-

6   fortunes

fortunes (for they are liable to many) fo that they grow fine trees, they will take up a great deal of room, for nothing will grow ftraight that their branches over-hang.

EITHER in natural or planted woods the trees may ftand fo thick as eight feet diftance, if the natural wood is managed with care after cutting, and the planted wood carefully pruned for a few years after planting.

THOSE that are fond of pleafure or pro-fit can have no profpect fo agreeable as a well-dreffed wood, where the trees are tall and ftraight, and not a foot of ground loft, the trees growing in a regular plea-fant form, which is agreeable even in win-ter to walk in, and in the fummer is a pleafant fhade ; at a diftance it is a beau-tiful carpet in the air, far furpaffing the moft elaborate works of what are called pleafure-grounds. All this may be accom-plifhed by following the directions before given.

I MAKE no doubt but it will be very
agreeable to all or moſt gentlemen to
blend pleaſure and profit, provided it can
be done without expence, and at a very
little loſs.

THERE are ſome trees that are of little
value, but very beautiful : if ſome ſmall
clumps of them were interſperſed in a
plantation with taſte, they would be very
ornamental, and the little room that ſhould
be allowed them would not be a great
loſs, as they would be of ſome value
when cut, and a great beauty when growing.

THIS may be done in all plantations,
even in the pooreſt barren ground ; for the
moſt beautiful of the flowering-trees will
thrive in the very pooreſt dry ground and
coldeſt ſituation.

IF the plantation is on a poor heath or
barren hill where there is little ſoil, there
may be planted clumps of hollies and
VOL. I.   M   tree-

tree-box ; for although they are not flowering-trees, they are beautiful ever-greens,
and will grow to a great fize in poor dry
grounds. The mountain afh and crab-
tree will alfo grow in fuch places : if thefe
four are properly planted they will be fufficient to add great beauty to a plantation
that is feen at a diftance.

The hollies fhould be planted at four
feet diftance to prevent their bufhing and
to make them grow tall ; there may be
ten of them in a clump.

The trees round them fhould be at ten
feet diftance at leaft, as they are flow growers for feveral years ; then at fome hundred yards, according to the largenefs of
the plantation, three or four crab-trees,
which fhould be planted at ten or twelve
feet diftance, that they may fpread and
have large heads. The trees round them
fhould not be too clofe to them, at leaft
ten feet diftance, as it would draw them
up and fpoil their beauty ; then fome
tree-

tree-box, which fhould be planted thick to make them grow ftraight and tall. None of the clumps fhould be planted in rows, but as irregular as poffible.

To beautify plantations on good land the fame trees fhould be recommended, with the addition of horfe-chefnut, common black cherry, filver and fpruce fir, (if no part of the plantation is planted with them). Plantations thus planted will make fine pictures at a diftance, and give joy to the beholder, and pleafure to the owner: they will have a chearful, agreeable look.

THESE clumps fhould not be too numerous, fo as to look like patches, but be planted at a great diftance, and placed fo as to add beauty to the whole foliage, for nothing is fo difagreeable, nor hurts elegance fo much, as plantations planted in little patches. Large clumps of a fort are noble and have a pleafant look: the light green of the box will add beauty to the

M 2   gloomy

gloomy fir; and the fhining dark green of the holly will make the fine green of the larch ftill more cheerful.

IF woods, hedges, and fields were managed as they ought to be, the whole country would be a delightful garden. The expence would be trifling confidering the advantage, for they would grow much better ; and as all gentlemen that are fond of real rural fcenes (and I believe moft are) fhould have the grounds in their own occupation in fuch order, that every wood fhould be a grove (inftead of a heap of rubbifh over-grown with thorns and briars) ; every grafs-field a lawn, only detached by a clean fallow, or a good crop of grain, to diverfify the fcene. And there might be fome art made ufe of, by decorations of evergreen, and detached trees and fhrubs at proper places, to add beauty to the whole.

EVEN in plantations that are converted into under-wood, where all ought to be

a thicket,

a thicket, there might be fome fmall or-
naments made round the outfides, fuch as
fmall clumps of hollies, laurels, and flowering fhrubs ; which may be fo placed as
to be feen by furprize, and not as if they
had been defignedly planted.

M 3   C H A P.

CHAP. IX.

On Fences, and their Management.

HEDGES are the moft ufeful as well as ornamental things belonging to a country ; but the method of managing them, in moft countries in England, is very erroneous. There is an old common tract of fcouring and laying them when they are grown fo bad as to be of little ufe.

THE method of planting them is much the fame, without confidering the nature of the ground, whether it is wet or dry, a ftrong clay, fand, or poor gravel. To have good hedges it would be right to plant them in very different ways.

THERE will be many objections made by the country-men in general, as it would put them out of their old way ; but let
any

any gentleman or farmer only try, and they will find it turn to their advantage.

BEFORE I begin to give directions for the planting the hedges, it will be neceffary to give fome inftructions concerning the raifing the plants.

THE berries are in general too foon gathered; they fhould not be pulled before the leaves are all off the trees. When they are gathered, they fhould be thrown into a heap, and lay for fix weeks; then they fhould be buried in a pit that is quite dry, and fhould have an inch of fifted coal-afhes in the bottom, and then a layer of berries, and then of afhes, until all is finifhed.

THE pit fhould be raifed a foot above the level of the ground, and a fmall trench made all round to prevent the wet getting into the pit. Coal-afhes is better than fand or mould to mix with the berries, as

M 4   it

it prevents all mouldinefs, and hinders mice from deftroying them, which often happens.

THE berries which are gathered when the leaves are on the trees, will grow, but as they are not then full ripe, many of them will mould and rot; and thofe that grow, will not make vigorous plants, which is their greateft qualification.

HERE they muft lay two winters and one fummer, until the time of fowing (the beginning of March) then they may be fifted from the afhes, or fown all to-gether, which is much the beft way.

THE ground fit for fowing them fhould be light and rich; and as there are many that do not come up before the fecond year, all that come up the firft year fhould be drawn in the winter; moft of them will be fit to plant out for hedges.

THE reafon for drawing out thofe that
3   come

come up the firſt year is, to let thoſe that come up the next have free air and room to grow. The ſeed-beds ſhould not be deſtroyed for three or four years, for if the large ones are drawn, thoſe that come up laſt may ſtand in the beds until they are fit for uſe.

EYERY time there are any drawn out of the ſeed-bed, it will be neceſſary to dig the alleys, and lay a little earth all over the beds, becauſe in the drawing there will be many of the ſeeds raiſed out of the ground.

IF quicks are wanted in a hurry, they may be made to come up the firſt year; that is, *many of them*, which will be fit for uſe next winter.

As ſoon as the deſigned quantity of haws are gathered, mix them well with twice as much freſh grains from the brewhouſe, (old grains will not anſwer) and lay them up in a round heap, covering them all round with half an inch of grains.

THERE they may remain for ten or twelve days, by which time they will be in a gentle heat ; they muſt then be turned and laid up as before, and lie as long, by which time moſt of the pulp will be rotted off; then they ſhould have another turning, and a good deal of ſand mixed with them, and rubbed between the hands, and laid up cloſe, and covered two inches thick with ſand to keep them from froſt. They muſt be under cover ; an open ſhed will do. Theſe muſt be ſown in March.

IF there is a hedge to plant in poor gravel or ſandy ſoil, pare off the graſs as thin as poſſible two feet wide ; take out a ſpade of earth and put in a quick, which ſhould be cut eight inches long, and laid in ſloping, ſo that the top ſhould not be above two inches out of the ground ; then dig on a foot and lay another in the ſame manner, and ſo on to the end; then there will be a ſtraight row at a foot diſtance.

OPEN

OPEN the ground two inches diftant from the firft row, and plant another quick in the fame manner between every one of thofe firft planted, fo that there will be another ftraight row two inches wide of the firft, and the quicks at fix inches diftance.

LET it grow at pleafure for two years, by which time the fhoots will be a good length; fo that they may be platted into one another, and then with a pair of garden fheers cut off all the long ftraggling branches. There is nothing further requifite for two years; then it fhould be platted, and cut as before.

IF the hedge is to grow tall, provide a hedge-bill with a long handle, and in winter fwitch off all the long fide-branches that over-hang the bottom of the hedge, and keep the top narrow, and it will be a good and neat fence for many years.

IN fuch grounds there is no occafion

for

for ditches, fo there will be no fcouring, and the dreffing will be a very fmall expence, as a man will fwitch and drefs as much in a day as he could fcour in a week, Scouring in fuch dry ground never ftands well.

In all poor gravel and fandy land, if a little very rotten dung can be fpared to put to the root of each quick, it would be of great fervice, and would encourage their growth greatly. If the hedge is to be kept clipt, which is the beft way to have a good fence on fuch poor land (for they grow very flow after laying) it fhould be clipped every year after the fecond platting. The expence of clipping will be lefs than where fcouring is wanted, and it may be performed in winter, when little other bufinefs can be done in the fields.

If the ground is ftrong clay there will be water to carry off, fo there muft be a ditch. The beft method to have a good

hedge

hedge is to make the ditch and throw the
beft of the earth up behind, and fo form
a border, and plant the hedge in the
border a foot from the face of the ditch,
in the fame manner as that planted on
poor land.

THE flope of the ditch on that fide the
quick is planted, fhould be faced with the
turf that is cut off the top of the ditch.
The turf fhould be a fpade deep, and the
firft fhould be laid as low as the bottom of
the ditch (if two inches lower it would be
better) to prevent the water undermining
it, which it frequently does where there
is a bare fpace left below the turf, fo that
the whole tumbles down.

THE advantages of planting the quicks
in this manner are many, and attended
with only one inconvenience ; that is,
there muft be a few thorns pricked in
the top of the flope, to prevent the cattle
fetting their feet on it, and to hinder
them from cropping the quicks and pulling down the flope.

IF the ground is good and no water to carry off, it would make a much better fence to plant on the level ground, as on poor land; for befides growing much better, there is no expence in cleaning and fcouring the ditches.

IT is three years longer in being a fence than if there was a ditch; but after it is fenceable, there is no further trouble than annually to fwitch off the long branches that over-hang the hedge bottom, and makes it foon naked when they are not taken off.

THE over-hanging of the top-branches kills all the bottoms; the hedge grows fo thin that it is not fenceable, and then it muft be laid; but by this annual trimming it will be very good for many years, and will grow much taller, and be a much better fhelter, where it is wanted for that ufe.

IF on good land there is water, and a ditch

ditch is neceffary, make it the fame as in clay land, and plant the quicks in the fame manner; and if turf can be got it would be right to face the flopes on both fides, as the light mould will tumble down with froft and rain and fill the ditch bottom foon.

I f the ground is wet, planting in the face of the flope is the beft method; but if it is boggy, quicks will not thrive; they grow cankered and ftunted, and never will make a fence.

I n boggy grounds Willows will make a good fence with little trouble. If a ditch will carry off the wet, make one as before directed; but if it will not, there is no occafion for any.

P r o v i d e good ftrong truncheons of the large growing kinds, three feet long; with an iron crow make holes two feet deep and eighteen inches diftant, and plant one into each hole, making it faft

2   with

with the foot. This may be done in any of the winter months.

The firſt year they will ſhoot many ſtrong ſhoots; let ſome of thoſe be platted a foot from the top of the truncheons, ſo then the fence will be two feet high : let all thoſe that are not platted in be cut off, but not too cloſe. The next year plat ſome of the ſtrongeſt ſhoots a foot higher, ſo that the hedge will then be three feet high.

Next year cut all the upright ſhoots a foot from the laſt platting, for there will be no occaſion for platting them any more. Thus there will be a fence four feet high which cannot be broke through. All the ſide-branches muſt be cut every year, and in a few years they will grow very thick and ſtrong, and be as good a fence as a quick hedge.

All hedges that are deſigned to be clipped ſhould be platted the third and
fourth

fourth year after planting, as it binds them fo that. they cannot be broke through ; but thofe that are to grow high, and are afterwards to be laid, fhould not be platted at all, for they would be fo intangled that it would be impoffible to feparate them.

THOSE on gravelly poor land make but bad fences when laid, fo that the beft way is to clip them ; but if that is thought too great an expence to the farmers, if they will carefully keep them thin at top, and cut off all the fide-branches that over-hang, they will laft at leaft twenty years and more, and be in a good condition to lay after that time.

IN good land and clay grounds laid hedges will anfwer very well. The cutting the boughs that hang down will be of great fervice to thofe that are to run up to height, and are to be cut down and laid. By the trimming of them annually

VOL. I.   N   they

they will remain good fences longer than can be imagined till tried.

HEDGES that are laid are in general spoiled in the bottom by weeds and grafs the firft and fecond year after laying, becaufe there are many root-weeds and grafs in the bottom of old hedges which have great roots, but have been prevented from growing in fummer by being fmothered by the hedges hanging over them; but when it is cut clofe they grow very faft, and almoft or quite cover the whole hedge, which is of great detriment to the young fhoots.

THE beft method to manage them is to lay the hedge firft, then with a ftubbing-hoe clean all the ground on both fides, taking up all the roots of briars, which run great lengths and do a deal of hurt; alfo all root-weeds and ftrong benty grafs; then face up the flope with turf from the ditch, always obferving to let the firft fpade of turf be laid even with the bot-

tom

tom of the ditch, and never to plaifter the mould up to a point againſt the quicks, but lay it in between and at leaſt a foot beyond the hedge. There ſhould be eight or ten inches flat between the ſlope and the hedge, which ſhould be kept as clean for two years as if it were a new-planted one, and by that time the hedge will be even with the face of the ſlope, and prevent weeds growing to hurt it.

In five or ſix years, if the ditch ſhould want cleaning and facing, give the turf that it is faced with a moderate ſlope; but by no means carry it nearer the hedge than it was at firſt.

Let the mould be broke down amongſt the roots of the hedge, and laid flat at top; this encourages its growth greatly; but when the mould is laid up to a point againſt the hedge, it deprives it of all moiſture; and when one ſide is covered up a foot, the other is quite bare. By being laid up in that way it is very

liable

liable to flide down, and is often much damaged by heavy fhowers.

I T is frequently difputed whether live or dead ftakes are beft in a laid hedge. I am of opinion dead ftakes are beft, for this reafon, that the live ftakes fhoot much ftronger and fafter than the branches that are laid, and grow quite upright.

I N a few years they over-hang the winders and kill them; and although they meet at top there are great gaps in the bottom. It will be of fervice, where there is any, to cut them clofe down to the binding of the hedge for two years, as it will encourage the growth of the winders, fo that the whole will be much clofer.

I F live ftakes could be left regular at two or three feet diftance, a good hedge might be made with a very little trouble. The winders fhould be laid in very thin, and all the fhoots that grow upright on

the

the top of the ftakes fhould be rubbed off
at leaft four times in a fummer. This
will caufe the ftakes to pufh many ftrong
fide-fhoots, which all grow almoft horizontally, and will foon meet.

To rub off the top-fhoots of the live
ftakes would be a trifling expence; a man
might go over a large farm in a day, if all
the hedges were laid in one year, which
never happens.

If hedges were managed as here directed, the expence would not be fo much
as it is in the common way, becaufe they
would laft much longer, and be much
better fences.

It is the common practice to plant trees
in hedges, and is recommended as beneficial; but it is an exceeding bad practice,
and fpoils the hedges, for they grow to
fhort boles and large heads; fo far as the
branches hang the fence is very weak, and
the trees are never of great value.

N 3   It

Iᴛ deſtroys and waſtes much ground, as little will grow under the drop of their branches, and the roots run a great way into the fields when in graſs, which are very troubleſome when taken into tillage.

Tʜᴇ beſt method is to plant a few trees in the corners of the fields, and where hedges meet. If all the four corners are planted, it will make a handſome clump; and, as the trees may be planted thick, they will grow ſtraight and tall, and make fine timber, which they never will do in hedge-rows.

Iғ the fields are large, the clump in each corner may be made ſo that when grown up it will be fine ſhelter for cattle in winter; and the fields may be kept ſeparate by a coarſe rail.

Tʀᴇᴇs thus planted will be more beautiful to the country, leſs damage to the farmers, and more profit to the landlords.

5   As

As they will be of courfe very irregular, from any eminence at a diftance the whole country will look like an entire wood, beautiful furpaffing imagination.

THERE is another improvement that would render the face of the country beautiful and fweet, viz. to plant red and white rofes in all hedges. Honey-fuckles are often planted, and are very fweet and pretty ; but they are fuch great runners, and grow fo thick, that they deftroy all the quicks, therefore I would not recommend them : fweet-briar will be full as fragrant, and be no detriment to the hedge. The common wood-rofe is pretty in hedges; but it grows fo vigoroufly that it is hurtful to the quicks. The red and white fpindle-tree might be planted in hedges that are to grow high; their berries are beautiful in winter : and if fome hollies were fcattered at a diftance they would look cheerful.

IT is a common practice (and as hedges
N 4   are

are managed there is no avoiding it) when
a hedge grows thin in the bottom, to draw
cut thorns into it to ſtop the gap; but in
reality it makes it larger, for the cut thorns
are drawn in very thick to ſerve the preſent
purpoſe, which kills the live wood, and in
a few years the whole bottom of the hedge
is ſtuck with dead thorns, when the top is
thick, ſtrong, and vigorous.

THE cutting up of the ſide-boughs that
hang down annually prevents all this; but
if by any accident there ſhould be an open
in the bottom of an hedge, ſtick in a few
ſmall ſtakes and wind ſome live branches
to them, and it will be cloſe in a year's
time, as the ſtakes will make the gap
good in the mean time, and be no detriment to the hedge.

CHAP.

C H A P. X.

On Vines.

THE cultivation of vines, and the bringing the fruit to maturity, has been much ftudied of late, and not without fuccefs; but, from a long and affiduous application, I have difcovered many things that are not in common practice, which I hope may be of fervice to the public.

I SHALL begin with raifing of the plants, and go through the management of them; firft, on common walls (for the benefit of thofe who refide in countries where they will ripen without help) fecondly, on fire-walls without glafs or covering, thirdly, on fire-walls with glafs or fome other covering; in glafs-houfes or ftoves built on purpofe for them,

them, called vineries, and in stoves that are used for pines.

I SHALL also recommend a method to prepare the borders for planting them, very different from what is in common practice, together with a particular method of obtaining good plants that will be very fruitful. I begin with the choosing of them, which is cuttings, they being much better than rooted plants, and, if managed as directed, will bear fruit sooner and much better.

THE method of procuring good vine cuttings has not been attended to with that attention which is requisite, as the chief part of the success depends on their being good. These should be taken from plants near the bottom of the wall, good bearing plants that have their eyes round and plump, their joints short, and the wood quite round.

CUTTINGS with these properties will be
very

very fruitful, whereas cuttings from the upper parts of the wall, although they may feem to have moft of the properties of thofe cut from the bottom, are far inferior, being in general longer jointed, the wood fofter, and more apt to fhoot into great rambling wood.

WHEN they are taken off, there fhould be an inch of the old wood to each, which fhould be cut floping. Three eyes are a fufficient length; the top fhould alfo be cut floping from the eye, and a quarter of an inch above it.

BEING thus prepared, put them into light dry ground (not too near together) up to the laft eye, preffing the ground clofe to them; and before the frofts come on, cover them over four inches thick with dried fern, or dried peafe-ftraw, and let them remain until you want them in fpring: they muft not be touched with the knife then, for it will caufe them to bleed, and fpoil them entirely.

SOME

Some may think that all this precaution is not neceſſary, and that a cutting taken from any part of the wall, if it has the properties (as to outward appearance) of thoſe taken from the bottom, is equally good; but let thoſe who think ſo only try the experiment, and they will find their error; the bottom cutting will ſhoot leſs vigorous wood, but be much more fruitful.

Cuttings are much the beſt for making all kinds of vine plantations. It is objected they are longer in bearing than young plants, which I never found to be the caſe; for if young plants are brought from any diſtance, it will be neceſſary to cut off all the ſmall fibres, which will ſo far impede their growth, that the ſhoots from the cuttings will be ſtronger, and bear fruit ſooner and much better.

Where there is an old plantation of vines, and a ſtrong plant is wanted for any particular place, if it is taken up carefully

with

with a good ball, and planted immediately, it will anfwer very well; but this
is only moving from one part of the garden to another a few plants; and if the
vine be very old, it is difficult to be done
with fuccefs.

THE common method of planting vines
(or cutting of vines, for there have been
fome walls planted with cuttings) is to
plant them at feven or eight feet diftance,
and fo train the fide-branches to fill the
wall; but there is a better way to have
finer grapes, and to have them bear fooner.

PREPARE fome good hot dung as for a
melon bed; make it up in the fame form,
but two feet wider than the frame that
is to cover it; two feet high will be fufficient, but it fhould be well trodden to prevent the heat being too ftrong at firft. Stake
it all round the fides with ftakes two feet
long, and wind them with ftraw ropes
very clofe to the top.

2   PRO-

PROCURE fome frefh tanner's bark from the tan-pits (if the bed is to be covered with a three-light melon frame, which will hold eighty plants) two good loads of bark will be fufficient, and fo in proportion. If the bark can be made dry before it is laid on the bed, it will be better; but as the weather is generally very uncertain at that time of the year (February) it may be dried in the bed.

LAY it on a ridge in the middle of the bed; fet on the frame and glaffes, and give it air: as the bark dries, draw it to the back and fore-fide of the frame till the whole is dried; then take off the frame, and fpread it all over the bed; fet on the frame and glaffes; let them lay clofe until the heat is rifen, which will be in three or four days, when it will be of a moderate temperature, if all has been performed as directed.

PREPARE fome mould a little lighter and richer than is generally ufed for melons.

lons.   Take penny garden-pots, and into each put one of the cuttings, but be sure there be an inch of mould between the bottom of the pot and the lower end of the cutting.

THERE should be only one eye left out of the mould, and the top of the cutting should be sloping to one side of the pot. When planted, plunge them into the tan half the depth of the pot at first (for they require little heat for some time) put on the glasses, and give them air night and day as long as there is any steam in the bed.

GIVE them no water while the steam is strong in the bed, for although they have air, they will be all wet with it in the night.   It will be proper to shade them in the day if it is very hot.

WHEN the buds begin to push (by which time the steam will be gone) give them a little water at the bottom when the

the mould appears dry ; but care muſt be taken to give it ſparingly, and often, for too much at a time would rot them. It will be of great ſervice to them to ſprinkle them all over twice a-week in the evening, and ſhut the glaſſes cloſe ; but they muſt have air in the morning early to dry the wet before the ſun becomes hot, for if they are wet it will ſcorch the end of the bud, and prevent its growing. They will require little water for the firſt three weeks.

WHEN the heat begins to decline, which will be in about five weeks (for the bark will grow mouldy, and then it loſes its heat) take all the pots out, and ſtir the bark down to the dung ; level it, and plunge the pots up to the top ; put on the glaſſes, and give a little air ; for although there will be neither heat nor ſteam for three or four days, there will be a rancid ſmell, which will be very prejudicial to the plants, and turn them yellow.

WHEN

WHEN the heat rifes, if it is moderate, which it generally is, there will be no further trouble with it, for it will laft long enough to bring the plants to perfection: but if it fhould be very hot, which fometimes happens, the pots muft be raifed until the violent heat is over, and then they may be let down, only ftirring the bark deep enough to plunge them to the top. They will now require a good deal of water and air, if the weather is warm.

IF they pufh more fhoots than one, take them off as foon as they appear; put a ftick to that which is to remain, and tie it up as it advances in height, for it will be of great detriment to the plant, and retard its growth, to let it lie down.

THE frame fhould be raifed as the plants grow high, fo as to keep the glaffes a foot from the top of the plants; but the bottom of the frame fhould be kept clofe, fo as no air can get in (which will be eafy to do, as the bed is broader than the frame)

VOL. I. O by

by sticking a few stakes round, and stuffing straw between the vacancies.

As the plants advance in height they will push out side-branches at the eyes, which must be constantly picked off as they grow; but not in the common method, which is to break them off close to the eye : they must be pinched off a joint above the eye, and then they will push again at that place, and never hurt the bud that is to remain for next year's shoot.

WHEN they are taken off close to the bud, it often pushes, and that joint is lost, which, if near the bottom, is of bad consequence; and although it does not push, there often comes one, sometimes two side-branches just by the eye, which weakens it greatly, and renders the place very unsightly.

THIS should be observed in taking off the side-shoots of vines of all ages, that they push again much sooner when they

they are pinched off at a joint than when
they are taken off clofe, and will require
a little more labour to keep them clean;
but by this method the eyes are all faved;
and as they pufh much weaker when they
are pinched off at a joint than when they
are taken off clofe, the vine buds for next
year are much ftronger, and confequently
the fitter for bearing good fruit.

THE bed for the vines fhould be made
the beginning of February; and, if all
fucceeds well, they will be fine ftrong
plants four and five feet high the beginning of June, at which time we fhould
begin to harden them; firft by taking off
the glaffes morning and evening, and giving them air all night; for if great care is
not taken, they will be ftopped in their
growth and turn red, which will be of
bad confequence, as then they will not
ftrike root after planting.

THERE is no other difficulty in keeping
them growing, than being cautious, giv-

ing

ing them air gradually, and not expofing them to the fun above an hour in the morning.

TAKE off the glaffes at five o'clock at night, let them ftand fo till ten o'clock, and after they have been ufed to that management for ten days, they may ftand uncovered all night, and have a great deal of air in the day-time. If a dull foft day happens they may be uncovered all day.

WHEN they are fo hardened that they can ftand the fun, they are then ready for planting, which will be about the beginning of July, and they will have time to ftrike root in the common borders before winter, and will be very fine ftrong plants.

THE common method of preparing borders for vines is to mix lime rubbifh and hard ftuff with good earth, which I have found, by many years· experience, to be wrong, and not a fit compofition for them. Borders in general for all kinds of fruit-

6

trees

trees are made too deep; two feet is quite
fufficient; the breadth ten feet at leaft.
When borders are made deep it encourages
the roots downwards, where neither fun
nor air can have any influence, and of confequence the fruit is not fo well flavoured.
If they are properly prepared there is no
danger from drought, which is the only
thing that can hurt fhallow borders.

In making of all kitchen-gardens. it is
a common practice to fet out the borders
and walks; then to empty the walks of all
good earth, and to fill them up with all
the rubbifh and ftones that are found in
making the garden.

The roots of moft trees run much further than is generally imagined, efpecially
vines; and when they meet with that bad
ftuff it cankers them, and infects the whole
tree. This is a good reafon why efpaliers, and other fruit-trees planted round
kitchen-gardens, fo often decay; for the

O 3　　　　　borders

borders are seldom above six feet broad,
and if they are planted in the middle they
have only three feet to run on the side next
to the walk before they reach this rubbish,
which cankers their roots and infects the
whole tree.

To make a good kitchen-garden, and to
give all the trees an equal chance, would
be to take levels of the whole; and, if it
can be conveniently done, it should slope
all to the south; but if the ground will
not admit of that, and some part falls to
the north, there should be a large covered
drain where the falls meet.

If the bottom is clay, it will be much
better to raise the level than dig into the
clay. None will attempt to make a kitchen-garden in a spot where there is not at
least eight or ten inches of tolerable good
mould.

It would be much better for all the
kinds of trees and herbage that are to
grow

in it, to let the clay, gravel, ftone, or
fand remain, and bring earth to make the
ground of a fufficient depth, as it will be
lefs expenfive; and there is no neceffity
that the eaft and weft walls fhould be level
at the top, nor that they fhould fall regularly to the bottom; for if the garden is
higheft in the middle, and falls to the right
and left, it will be no offence to the eye,
nor detriment to the ground, provided
there be a fall to the fouth.

I F any little heights intervene they
fhould be taken down, fo as to make the
bottom of the ground have the fame level
with the top; for that is of more confequence than is generally imagined. If the
bottom be clay, and there is a hollow dug,
it holds water; and when the roots of
trees reach the place, they are rotted.

A HEIGHT has the fame effect, as it
ftops the water, throws it back, and makes
the ground all round very wet.

HILLS

HILLS of ſtone or ſand are alſo of bad conſequence, for the roots are either ſtopped by them, or they run into them, which is much worſe. If the bottom has no obſtructions the water paſſes off regularly, and the roots of trees and plants meet with no impediment under ground, ſo that all things will thrive well.

WHEN the levels are fixed upon, the walks ſhould be ſtaked out and dug over, and cleared of all weeds, ſtones, and roots, as deep as the bottom of the other parts of the garden, and left ſix inches lower than the level, which will be ſufficient to lay gravel, or any other kind of hard ſtuff, to form the walks, for graſs walks are very unfit in kitchen-gardens.

BUT if any gentleman prefers them to gravel, they ſhould be left only two inches lower than the kitchen ground and borders, which ſhould be filled up with ſharp ſand to lay the turf upon, as it will make them much drier for walking on in winter, will

will greatly prevent the worms from working in them, and keep them from growing
too flufh in fummer, which they would
do, as the ground is all made good below.
They fhould alfo be inade a little round to
make the wet run off.

If the kitchen-garden is a fmall one,
the walks fhould be narrow in proportion;
in that cafe the whole ground may be
trenched, and the walks emptied either
for grafs or gravel, as the owner choofes,
and the foil that is taken out of the walks
may be fcattered over the kitchen ground.

If the walks are thus formed and made
good, the borders may be contracted to
what breadth the owner pleafes, and it
would be no detriment to the trees on the
walls, if the gravel was laid clofe to the
wall.

But it would be fo far a lofs, that
there would be no fouth borders for planting things to ftand the winter; and every
five

five or fix years it would be neceffary to
take up the gravel to add fome frefh compofition for the encouragement of the trees,
which fhould be prepared fome time before it is ufed.   By this management the
trees will thrive very well, and the garden
will be exceeding neat.

If the gravel is laid clofe to the wall, or
the borders contracted, there fhould be at
leaft ten feet diftance from the wall, made
as good, and in the fame manner, as if the
borders were of that breadth, which is
fufficient for the roots, provided they meet
with no obftruction after they have run
that length.

This digreffion is for the advantage of
thofe who are making new kitchen-gardens, as it will be of great fervice to efpaliers and dwarf apple-trees on the fides
of the walks, and of great ufe to the wall
trees; for although the borders are twelve
feet broad, which is wider than can be

2   afforded

afforded in a fmall kitchen-garden, befides it is in no proportion, and very un-
feemly to fee a fhort narrow walk and fo
broad a border; and if the trees are fo
planted that the roots be made to run horizontally, they in a few years will reach
the fide of the walks, and then the roots
are liable to all the misfortunes that they
would have been, had the border been
narrower; only then it would have happened fooner, which any one may fee that
will be at the trouble to examine into the
bottom of a rubbifhy gravel walk, although the border was fourteen feet broad.

If ever there have been thriving trees on
the wall, the roots will be found caukered
and full of knobs and bunches. I do
not pretend to fay it is from that reafon
that trees do not thrive; they are liable to
many other misfortunes, and often decay
before their roots extend half way over
the border: but it is my opinion that
moft of the thriving trees off fix and feven years old, that go of by canker, are
infected from that caufe.

I AM confident that thofe who pleafe to try this method of making their kitchen-gardens, will find their trees flourifh much better than by any method that has hitherto been practifed; and the widenefs of the borders, which has been introduced of late years, is a certain reafon that what is here directed will be of greater fervice. But to return to the making the borders for Vines.

IF the natural foil is a light fandy loam (which is the beft) with a clay bottom, raife the ground fo as to have a fufficient depth (two feet) which is much better than digging down into the clay, although it fhould be neceffary to have a ftep or two into the garden.

SPREAD fix inches of rotten dung all over the top of the border; open a trench two feet deep; lay three inches of the dung in the bottom; and when the border is a foot high, lay in the other three inches, and then fill up the border to its level;

level; leave the laſt trench open, and in about three weeks work it back again; only when within four inches of the dung that is lain in the bottom, it ſhould be dug over, and the dung and the ſoil well mixed. The dung that is laid in the middle will mix in the courſe of working, but none of it ſhould be brought to the top. This laſt trenching ſhould be at leaſt ſix weeks before the border is planted, in order that the ground may be well ſettled.

HALF a year at leaſt before planting make up the following compoſition, the quantity according to the number of vines that are to be planted: Two loads of good freſh loam from a paſture, two loads of rotten wood that is become earth, one load of ſharp ſand, and one load of very rotten dung; lay it in a long ridge four feet broad at bottom, and quite narrow at top: this muſt be turned often, until it is ſo well mixed that none of the ingredients can be diſtinguiſhed. If this

was

was made up in the beginning of winter, and turned over when hard frozen (without fnow) it would be of very great fer- to it.

THE borders, compofition, and plants being now all ready, provide flat flags or ftrong flate a foot fquare. For as many plants as the plantation confifts of, make holes a foot fquare, one foot deep, and at two feet diftance; lay the flags or flates into the holes, fcatter the mould that was taken out of the holes all over the border, and lay an equal quantity of the prepared compoft down at every hole; put into the hole, upon the flag, fo much of the com- pofition as that, when the plant is put into it, the ball may be covered an inch.

TAKE the plant carefully out of the pot, loofening the roots gently; but great care muft be taken not to break them; then place them in the hole eight inches from the wall, their heads inclining to it; fill up the whole with the compofition all round

round the ball, as high as the top, preſſing it gently with the hand, and ſo proceed until the whole is planted; then give
them a good watering, and a little after
cover the ball an inch over the top with
the compoſition; faſten the plant to the
wall with two or three nails, according to
its length; ſpread an inch of rotten dung
all over the border, and prick it in four
inches deep.

IF all has been managed as directed they
will grow a good deal after being planted,
therefore muſt be faſtened to the wall, and
have the ſide-ſhoots picked off as they advance in height and puſh out; but as the
lower part of the vine will have given over
puſhing ſide-ſhoots, thoſe now at the top
may be broke cloſe off, as all that part will
be cut off in pruning, but is now allowed
to grow to ſtrengthen the lower part of the
vine, which grows much ſtronger than if
they were ſtopped at the top; for when
they are, they puſh many vigorous ſide-
ſhoots.

THE

THE ufe of the flag is to turn the roots horizontally at their firft growing; and as the plants have been raifed in pots, the roots are prevented from taking a downright pofition; and although the bottom of the border is hard clay, there will be no occafion for any rubbifh or ftones being laid in it, for the roots will run horizontally, and never attempt going into the clay.

IF cuttings are taken off and preferved, as before directed, during the winter, and planted, as above, againft either common or hot walls, they will anfwer very well. They fhould be planted in the beginning of March; they will do much better than thofe that are raifed in a feparate place, and taken up and planted againft the walls afterwards; but they will be two years longer in bearing fruit than thofe that are raifed and trained in heat, and will not be fo ftrong the third year as thofe raifed in heat will be the firft. Vines are much better never to be removed.

ABOUT

ABOUT the beginning of Auguft nip off the end of the leading fhoot; for if it is done fooner they will pufh out fo many fide-branches, that it will weaken the plants greatly, and they will ceafe growing.

As foon as the leaves grow red, which will be about the end of September, prune them, and moft of them will be ftrong; cut them, according to their ftrength, into one, two, or three eyes; let the flope of the cut be oppofite to the eye, and a quarter of an inch above it; and as you cut them have fome clay at hand (a little fofter than for graffing) and over every cut give a very thin cover; this will be of great fervice to them; for the wood of young Vines is foft and fpongy, and their hearts opener than thofe that are come to maturity.

THE thin coat of clay prevents wet and froft from penetrating them; and being cut a little above the eye, although the froft be fevere, it never damages the bud, which often happens when cut clofe to it.

IF the plants raifed in pots are planted againft a flued wall (for they will do very well on common walls, in countries where they ripen without affiftance) it will be of great fervice to add a little fire to them for two months after they are planted, as it would encourage their growth greatly, and ripen the wood much better.

THE border which has been prepared and planted is fuppofed to be a fandy loam, which is certainly the beft; but as that cannot always be had, it will be neceffary to give fome directions for preparing all the different foils which can happen, at leaft thofe that are fit to plant Vines in.

IF it is a ftrong loam, which always has a hard clay bottom, but in general a good depth of foil, a compofition of light black

black earth, ſharp ſand, and coal or wood
aſhes (if coal aſhes, they muſt be ſifted
through a fine riddle, as any pieces of
coal would canker) and it will make the
border in good order.

IF the ground is- in graſs, and a ſtrong
loam, open a trench two feet deep, pare off
the graſs two inches thick, and lay it in the
bottom of the open trench; then two
inches of rotten dung; then ſix inches of
the natural earth, two inches of the compoſition, four inches of natural earth, two
inches of dung, two inches of natural earth,
and four inches of the compoſition; let
all be well mixed (only the turf at the
bottom ſhould not be diſturbed) to incorporate the different ſtrata, and they will
then be fit for planting.

IF the ground is ſandy, or a light black
earth (the ſame materials will anſwer for
both) a good quantity of a ſtrong loam and
rotten dung, the dung and loam equal to

P 2

a third

a third of the natural foil, and well mixed, will make it in good order for planting.

A HARD dry gravel is thought the propereft foil for Vines of any, and not without good reafon, as moft of the fineft wines are produced from fuch grounds; and I have been informed that the fineft of the Spanifh wines are produced from the grapes that grow wild on the rocks without any cultivation; and from thofe obfervations it was recommended to mix lime rubbifh and gravel with the foil on borders that are intended for Vines, and the directions feem very rational; but experience and the obfervations of many years have made it plainly appear to me to be a wrong practice.

THE grapes in thofe countries are fmall, and the great heat brings them to perfection; but even there thofe forts are not efteemed for eating: befides, they are not allowed to grow above three or four feet high,

high, and to carry a few clufters on each
branch, which no ways anfwers clothing
a wall of ten or twelve feet high from top
to bottom; the wood, foliage, and fruit
muft be fupported to bring them to perfection; and I have found, by long practice,
that Vines thrive, and carry large and good-
flavoured fruit, when planted in borders
prepared as here directed.

I HAVE tried many compofitions, and
all hard ftuffs, fuch as gravel, rubbifh, &c.
and they never produced fo good-flavoured
fruit. I have been informed that all the
vineyards round Rome are on a fine rich
deep loam, and are manured with the afhes
and the cleanings of the ftreets, and that
they produce the beft of eating grapes. It
from this hint I firft began to alter the
preparation of the borders for Vines,
which has been crowned with many years
fuccefs.

IF the ground is a hard gravel, add four
loads of rotten dung to fixteen loads of
good

good ſtrong loam and eight loads of the
natural ſoil; mix them well, and after the
border has been worked all over in that
manner it will be fit for planting.

IN all the different ſoils that have been
treated of, the ſame manner of planting
muſt be obſerved that is directed in the
firſt; for the goodneſs of all fruit depends
greatly on having the roots run horizontally and ſhallow, and none more than
Vines.

IF the cuttings raiſed in heat are deſigned
for a hot-houſe adapted for Vines, or a
ſtove that is employed for Pines, the beſt
way would be to plant them as ſoon as
they are fit, for it is impoſſible to harden
any thing that is brought up under glaſſes,
ſo as to make it endure the open air without giving it a check; and as they are
both covered with glaſſes, the plants will
ſuffer nothing by being planted as ſoon as
they are ready.

THOSE

THOSE in the hot-houfe that are for Pines will thrive and grow beyond what many will imagine; and thofe that are planted where there are no Pines, if they are managed in the fame manner, will anfwer beyond the moft fanguine expectation. The preparation of the ground for them fhall be treated of when we come to give directions for the management of Vines in in ftoves and vineries.

GREAT care muft be taken the next fpring, when they begin to fhoot, to keep all fhoots that come from the bottom rubbed off as foon as they appear, and none allowed to grow but thofe from the buds left in pruning. They fhould be carefully kept to the wall, and all the fide-branches nipped off, and allowed to run to the end of July; they fhould then be ftopped. There will be fome little clufters of grapes this year; but the encouragement of the Vines is the only thing to be regarded, fo that there fhould be moderate fires kept.

I FOUND

I found by the experience of many years that moderate fires, which would bring the fruit to perfection in September, are a great advantage to young vines, as they ripen the wood, and make them fit for pruning early in autumn.

Young plants that are vigorous grow much longer than old ones, and when they have no affiftance they are not fit to cut till late in the feafon, which is of very bad confequence to them ; for they fhould not be pruned at a time when there is a probability of froft, but it fhould not be deferred till fpring; for although they are late in fhooting, if they are pruned any time in March, which is as foon as it can be done with fafety, and at that time they have no appearance of vegetation, yet as foon as they begin to pufh (and fometimes before) they will bleed much, although every gardener knows how to ftop their bleeding, and may prevent their receiving any damage ; yet their bleeding, when cut in fpring, fhews evidently it is wrong,

as

as they never bleed when cut in the autumn.

IF the wood is not fufficiently ripened by the end of July (which it generally is) keep on the fires ten days longer; then let them out gradually, and they will be fit to prune in the middle of September, without danger, as the frofts feldom are very hard at that feafon.

THE Vines will now be ftrong and vigorous, and, at this cutting, every fixth plant may be left a yard long, and thofe between to three, four, and five eyes, according to their ftrength.

IN order to get the wall covered with bearing wood, cut them floping as before directed, and nail them; then fpread fome very rotten dung four feet broad from the wall, and prick it in with a dung fork (a fpade fhould never be ufed nearer than four feet of the wall); then all is finifhed for this feafon.

THE

THE next fpring, about the end of March, make on the fires, flow at firft, but increafe them gradually, fo that when the Vines have fhot about an inch the walls fhould be milk-warm all over, and fhould never be much hotter.

RUB off all branches that attempt to fhoot from any part of the Vine, except from the eyes left in pruning; carefully nip off the fide-branches, and keep the young fhoots nailed as they grow.

THERE will this year be a tolerably good crop of grapes; and as foon as they can be diftinguifhed, mark the wood for next year, which fhould be the fhoot next the old wood; and when thofe that have fruit, and are not for next year's wood, are grown two or three joints beyond the fruit, top them; but thofe that are defigned for the next year's wood, although they have fruit, muft not be ftopped before the latter end of July, which will caufe them to grow ftrong and vigorous for next year's crop.

THE fruit on them will be smaller and later than those that were stopped, but they will come to succeed the others, and it will be a great advantage to the wood.

WHEN the fruit is out of blossom, and as large as a pin's head, if the weather is dry and warm, they should be watered twice a week; but if there be showers, once a week will be sufficient, until the fruit is come to its size, after which it should have no more water at bottom; but if the weather is very dry, and there is little dew, if the whole wall was sprinkled in the evening, once a week, it would increase the size of the fruit.

THERE is nothing further requisite until pruning time, but to keep them clean of all shoots that attempt to shoot from the old wood and the side-branches, which must never be neglected.

IT would be of great advantage to both

fruit

fruit and Vines, at the beginning of the year to lay some short grass close to the wall, four feet broad and two inches thick. This will keep them moist after watering, and prevent the ground from cracking, which frequent watering occasions; for if it cracks, the roots of the Vines being near the surface of the ground, it would destroy all the small fibrous roots, and greatly damage the fruit, and retard their growth.

When the grapes are all shewn, and the next year's wood fixed on, if there are any shoots that have no fruit, pull them off. If they come clean off (as they should) they will leave a small hole in the old wood, which should be immediately filled up with worked clay; for very soon after the shoot is pulled off it begins to bleed, after which it will be difficult to get the clay to stick; if it bleeds, it will hurt all the fruit on that branch, and greatly weaken the Vine.

By

By this method there will be no ufelefs wood on the wall to weaken the tree, nor to croud the branches that bear fruit, or thofe that are for next year's wood, which is often the cafe.

Every pruning feafon fome of thofe that have long bare wood may be cut down; fo that by cutting fome down every pruning, the bottom of the wall may be kept as well covered with ftrong wood as any part of it. A great quantity of good cuttings cannot be had from a well-managed wall; but when many plants are wanted, if the fide-branches of thofe that bear fruit are nipped off at a joint, as directed for the wood of next year, there may be a good many got.

The fruit being all gathered, and the pruning feafon come, the loweft fhoot on the plant of all thofe Vines that were left fix feet long laft pruning fhould be cut to four eyes long, to keep the bottom of the wall in good bearing wood; and the top

5   fhoot

fhoot of the fame plant to five or fix eyes,
to fill the middle of the wall with young
wood; and all the fhoots between the top
and bottom fhoot cut off clofe to the laft
year's wood. There fhould alfo, at this
pruning, be fome more fhoots, fix feet long,
laid in exactly in the middle between thofe
before left.

ALL the other branches fhould be cut
to five and fix eyes, according to their
ftrength; obferving that a weak fhoot
fhould never be left with more than three
eyes.

As foon as the Vines are pruned, let
them be nailed, the border cleaned, and the
ground forked out four feet broad from the
wall; the other part of the border dunged
two inches thick with very rotten dung
that has been turned feveral times, and is
very mellow: this fhould be dug in with
the fpade.

WHEN the Vines begin to fhoot next
3   fpring,

spring, great care muft be taken to rub off all fhoots fpringing from the old wood; for thofe fhoots that are laid in long, and had the branches cut clofe to the old wood, will pufh out many fhoots at every amputation, and many will alfo fpring from other places on the old wood, which muft never be fuffered to grow.

THE keeping clean, and pinching off the fide-branches, is every year the fame, fo need no more to be repeated.

ON the proper difpofition of the wood at this feafon depends the beauty and regularity of next year; fo that great attention fhould be had to the following directions.

THERE being now on the wall two forts of long fhoots, one two years old, which has a young fhoot of five eyes at bottom, and another much of the fame length at top, the eye next the old wood fhould be marked for next year's wood on the low fhoot,

fhoot, to fupply the bottom of the wall;
and the fifth or laft eye on the top fhoot,
to furnifh the top of the wall.

THE laft laid in fhoot, which is all
young wood, muft have the loweft and
higheft eyes encouraged for next year; and
all the other plants on the wall fhould have
the eye next the old wood trained for next
feafon.

THERE will be fome fhoots that have
no fruit on them; if they are not marked
for next year's wood, they fhould be
pulled off to give air to the fruit and
wood: there will be a hole in the old
Vine where the fhoot came out, which
muft inftantly be filled with well-worked
clay; for if not ftopped it will foon begin
to bleed, after which it will be difficult
to make the clay ftick. If it be allowed
to bleed it weakens the Vine and fpoils the
fruit.

ALL the branches that have fruit, and
are

are not for next year's wood, fhould be
ftopped as foon as they have run two eyes
beyond the laft clufter; thofe for wood
fhould not be ftopped before the end of
July.

THE next pruning the wall will be full
of good bearing wood, fo that it fhould
be left regular, and the fhoots cut to four,
five, and fix eyes, according to their
ftrength. There fhould this feafon be fome
rotten dung fpread four feet broad from
the wall, and forked as before directed.

IF Vines are thriving, and carry great
crops of fruit, they will only require
dung every other year after the fourth;
but it will be better to dung one year
half the breadth of the border next the
Vines; then mifs a year, and dung the
other half.

THERE is now only one difficulty remaining in the management of Vines on
walls, and that is to keep the bottom of
VOL. I.   Q   the

the wall full of good bearing wood, which
it is impoſſible to do if the Vines are
planted at a great diſtance.

THERE was a neceſſity to run ſome
ſhoots a great length, to furniſh the top of
the wall until the other wood mounted
gradually, which it is now ſuppoſed to
have done.

THOSE branches that were laid in ſix
feet long, and at ſix feet diſtance, having
one ſhoot at bottom and another at top
ſaved for wood, and of conſequence the
middle being now all bare, cut them down
to the ground; thoſe cut down will puſh
out many ſhoots from the bottom, which
muſt be all pulled up but one of the
ſtrongeſt on each plant, and it will be a
fine ſtrong ſhoot for next year.

THE next pruning ſeaſon there ſhould
be ſome more of the long ſhoots cut down,
and managed in the ſame way as in the
others; and every year there will be ſome
old

old plants that may be cut down, fo that the wall may be kept in good bearing wood at the bottom for many years. There is no difficulty in keeping the middle and top of the wall in young wood.

THERE cannot be many good cuttings obtained from a Vine-wall that is properly managed; but if plants are wanted, pinch off the fide-branches in the fame manner from the fruit-bearing branches as they are from the branches that are defigned for next year's wood, and there will be more good plants; but unlefs the plants are wanted, the beft way is to break off the fide-fhoots clofe to the eyes of thofe that are not for wood: they do not fhoot fo foon as when pinched off at a joint above the eye, fo require lefs labour.

IT may be thought that more dung is recommended than neceffary, and that it will fpoil the flavour of the fruit; but if rightly confidered, there muft certainly be a great fupply to fupport plants on a wall

Q 2   twelve

twelve feet high, covered with luxuriant branches, fruit, and fuch great quantities of leaves as there are on a well-managed Vine-wall.

WHERE this is not obferved, the wood is very fmall, the clufters little, and the fkin of the fruit very tough, and of a very bad flavour; but where they are properly fupplied with rotten dung that has been prepared at leaft a twelvemonth, they will continue many years in the greateft vigour. By this management I forced one wall twenty years, and had always good crops of well-flavoured fruit, large clufters, and the Vines in good order.

BY the common method of planting Vines at a great diftance, and training the fide-branches until they meet, in a few years the bottom of the wall becomes bare, without a poffibility of preventing it; but by planting at two feet diftance there is a command of bottom-wood (if managed as directed) for many years.

THIS

THIS method of planting, training, and pruning of Vines will anfwer very well on common walls in the fouth, where they ripen without fire.

THE management of Vines on a fire-wall with glaffes, as to preparing the borders, planting, dreffing, and pruning, is the fame as before directed for walls without glaffes.

THE management of the glaffes is the thing now to be confidered. They fhould be made to take away at pleafure, and never fhould be ufed until the Vines are of a good ftrength, which will be the fourth year, if planted in the common method; but if from cuttings raifed in heat, they will be fit the fecond year, if managed as follows.

As foon as the plants are fit for plant-ing, without being hardened, plant them on the fire-wall; put on the glaffes and a gentle fire; if they have fucceeded well

Q 3

in

in the raifing they will be fit to plant
the beginning of June. Give them moderate heat and air, fo as not to draw
them, and they will be fine ftrong plants
by the end of Auguft, when the fires fhould
be put out, and more air given to harden
them.

If they have all thriven they will be very
ftrong. As foon as the wood is well hardened they fhould be pruned, which may
be done about the end of September.

They muft be cut very different from
any thing yet mentioned. Begin at an end,
cut the firft to four eyes, the next to ten,
and fo on until all are finifhed; nail them
immediately, and drefs the border.

The reafon for this difference in cutting
thofe young trees is, that if they were all
cut fhort there would be little or no fruit;
and if they were all cut long, there would
be fo much wood that the wall would be
crouded, and not have room fufficient to
lay

lay in the wood, fo as to have proper air, which would fpoil the next year's crop; but by this management there will be a good deal of fruit, and the Vines remain in good order for the next year: and here it will be abfolutely neceffary to pull off all barren branches that are not for next year's wood, as there fhould be nothing but what is ufeful left to weaken the plants.

It will not be proper to force them early in the fpring, as they will not then be ftrong enough; but if the fire is put to them the beginning of April, and kept on that month before the glaffes are ufed, it will be much better for them: when the glaffes are put on, give moderate heat and air, and there will be a tolerably good crop of grapes, which will be ripe the latter end of July.

It will do the Vines no hurt by forcing them fo young, provided it is done moderately; and as this year the wood will be

ripened

ripened much better and fooner, they may
be pruned early, and fo have the heat applied fooner in the fpring.

THOSE who delight to have grapes very
early, fhould begin in January to apply the
heat; for it is much better to begin early
and work moderately, than to keep them
very hot, as it does lefs damage to the
Vines, and there is more certainty of a
crop; but it is impoffible to force in fo
much bad weather as generally happens in
thofe cold months, without hurting the
plants; fo that thofe who choofe to have
them early in the feafon, fhould have two
feparate lengths for that ufe. Thofe which
were forced early this year fhould have no
glaffes put on them the next year, but
fhould have a moderate fire applied the
beginning of April, and continued until
the end of July.

IN the pruning feafon, that which was
forced early fhould be cut fhort; that is,
have few eyes left on a fhoot, in order that
                                        they

they may pufh ftrong wood for next year,
as the preparation for the next early crop
is what is to be regarded this feafon.
Thofe who prefer a good crop to having
them very early, may force the fame wall
with glaffes for many years with good
fuccefs.

THE beginning of March is a good feafon to begin to apply the heat, if the
weather will permit. It is much better
to have the fire at the wall ten or fifteen
days before the glaffes are put on; but it
will be dangerous to keep them long without glafs at that feafon; for after they
begin to pufh, and the fhoot is juft coming
out, if the froft fhould catch the top of it,
all hopes of a crop are fruftrated for the
feafon.

GIVE a good deal of air every day,
efpecially when the fun is hot; for if
drawn, they drop their fruit after it is
formed, before they come into bloffom.
There fhould be little or no water given
until

until the fruit is out of bloſſom, and then
they ſhould frequently have a little; but
at firſt it ſhould be given in the morning,
when there is the appearance of a fine day
and little wind. As ſoon as they are
watered, the glaſſes ſhould be opened a
good deal to dry the ſurface of the ground
before night; for when the glaſſes are ſhut,
there is a great ſteam, which is very prejudicial to the young fruit, and often makes
it drop off; but when half grown, it is of
great ſervice, and then they ſhould be
watered in the evening, and the glaſſes
ſhut immediately.

WHEN the fruit is as large as a ſmall
pea they will bear more heat, and the
glaſſes need not be opened ſo ſoon in the
morning, and may be ſhut cloſe at night
before they loſe the ſun; but in the middle
of the day, if the ſun is hot, they ſhould
have a conſiderable quantity of air. In
dull days a little will be ſufficient; but if
it is even a cold day, they ſhould have a
little to rarefy the air, otherwiſe the fruit
will not be good,

WHEN the grapes are come to their fize they may be kept pretty hot and clofe, if they are wanted very early; by that management they will grow very large and fine to look at, but will not be fo high-flavoured as when they have lefs heat and more air.

WHEN the fruit is in bloffom they fhould have no water; and the place kept as dry as poffible, for they are then in the moft critical fituation, as they at that time receive great quantitie of the fteam (if there is any) that rifes when the glaffes are fhut; fo if great care is not taken, many of them will drop off; to prevent which, if two or three little holes were made in the back-wall with fhutters, they might be opened in the night when the fruit was in bloffom, and it would keep the place quite dry

WHEN glaffes are ufed for grapes on fire-walls, and the fruit is not required early, there fhould be two lengths, which will be be fufficient to make a regular fucceffion, The frames and glaffes fhould be moveable; and thofe that were forced with glaffes, if properly managed according to what has been directed, will be little worfe.

As foon as the fruit is gone the glaffes fhould be removed, and the Vines cut, which will be about the beginning of September, when they fhould be nailed, and the border dreffed.

THOSE Vines that had glaffes over them laft year, and are to be this year without, fhould have fire put to them any time between the 25th of March and the 1ft of May, and in every refpect managed as directed for fire-walls without glaffes.

BY this means there will be as good a crop this year as if there had been no glaffes the year before, and the Vines will be very fit for glaffes the next fpring; and thofe which are forced by fire only this year,

year, will gain ftrength if they were any
ways drawn : they will alfo fucceed thofe
under the glaffes, and there will be a fufficient quantity to fupply a moderate family from the middle of June to the end
of October.

I have had feveral crops of fine grapes
by the following method : I applied the
fire to the wall the 1ft of April, and worked
it to the end of May without glaffes, in
the fame manner as if there had been
no glaffes intended. At that time, the
grapes being juft out of bloffom, I put on
the glaffes, and managed them in the fame
manner as I have directed for Vines under
glafs, and had very fine crops. The berries were very large, the fruit had a much
finer flavour than thofe that had the glaffes
put over them fooner, and were ripe the
latter end of July.

Proper heat and air are the only things
to be confidered to render fruit of a good
flavour ; and unlefs thefe two elements
are ufed properly, grapes will never be
good :

good: they may be obtained early by the ftrength of heat, and by a large quantity of water they may be fwelled to a large fize, but they will be infipid.

A FINE large clufter of grapes makes a fine appearance when the berries are plump and large; but in eating they tafte flat, and have no flavour, which is a great difappointment; but thofe who follow the directions here given with accuracy, will obtain fruit rich and good, which cannot fail to give both fatisfaction and pleafure.

THE directions for the management of grapes on fire-walls forced early, and thofe forced late, may feem much the fame; but when compared they will be found very different. The wood of the firft will be much weakened by having early fruit, which it is impoffible to prevent, as the inclemency of the weather in the winter months will not permit giving them fufficient air to keep the wood ftrong; fo that the management of them next feafon (although fire is requifite) is not with

regard

regard to fruit, but to recruit the wood, and make it fit for forcing early the next year with glaffes ; whereas thofe that are not forced early will have a good crop under the glaffes, for at that feafon proper air maybe given to keep the wood in good order.

THE year following there may be as good a crop as if there had been no glafs over them the preceding year. As to the forcing of the young Vines that were raifed in the hot-bed of tanner's bark, it was only to encourage their growth, as it is directed to be done with moderation, although, at the fame time, there will be a good deal of fruit.

GRAPES may be forced with paper co-vers inftead of glafs ; and if the wall be lighted the beginning of April, and worked, as if no covers were intended, to the beginning or middle of May, they will anfwer very well and bring the fruit forward ; but when they are ufed early they are liable to many misfortunes from the weather ; and in a ftormy night of wind and froft, if the covers are torn (which fometimes happens) much damage may be done to the crop.

FRUIT raifed under paper covers are not fo high-flavoured as when under glafs, and by the faintnefs of their light they draw and tender the trees much more, efpecially in dull days, when much air cannot be given.

HAVING treated fully of fire-walls with and without glaffes, I now proceed to hot-houfes built on purpofe for vines, called Vineries.

THERE are many forts and forms of Vineries which bring good crops of fruit. Some of them have the Vines planted on borders in the houfe, and many are planted on the outfide, and taken in at holes left for that purpofe.

IF the houfe is intended to have grapes very early, it is beft to have the Vines planted on a border in the houfe : if for a large crop, and of a good flavour, they are beft planted on a border on the outfide,

3   for

for there the roots have the advantage of fun, air, and natural rains, which is much better than any artificial waterings, although ever fo fkilfully performed.

A VINERY built on purpofe fhould have no tan-pit, although many hold it beft, as it gives a natural moift heat, which is by fome faid to be better than the dry fcorching heat produced by fire.

A SOFT moift heat is beft for all kinds of fruit in the open air; but is not fo where the air is confined, for many reafons too numerous to mention, fo fhall only take notice of two.

IF there fhould be a great heat in the bark when the vines are in bloffom, the fteam that rifes from it caufes many of them to drop off. When the bark has loft its heat at top, which it foon does, it muft be often ftirred up, which caufes a very bad fmell, and a great fteam rifes, if it is frefh and hot; if there is not much heat, it has a mufty fmell, which will infect the fruit, unlefs there are air-holes to

VOL. I. R carry

carry off the fteam at night when the houfe is fhut up ; if that be the cafe, the moift heat of the bark is loft, therefore would be better without it.

A good Vinery fhould be all flagged, and the flues fo conftructed that they may give a good deal of heat with little fire. There fhould be a flue all round the houfe, which fhould ftand above the level of the floor, and as there is no tan-pit, it fhould ftand off the walls at leaft two feet, that the heat may be equal on both fides, which it cannot be when clofe to, or in the wall. There might be a fingle flue all along the middle of the floor, and one in the back wall, independent of each other. A little fire will keep a houfe warm if thus conftructed.

A house built after this method might have the border made between the flue and the forefide of the houfe thus : A brick of breadth laid a foot from the flue and raifed one foot, the other foot funk ; fo that the top of the flue, which fhould be

two

two feet high, would be a foot higher than
the surface of the border. The foreside
wall of the house should be built on
arches supported by small pillars (which
there would be no difficulty in doing, as
there is no flue in the foreside) so that the
roots of the Vines might have liberty to
strike into the border on the outside,
which might be made ten feet broad.
The arches should be all so low that their
tops may be within the border.

The Vines should be planted on the inside of the border, a foot from the brick
that supports the mould, by which means
the roots would soon be without the house,
and have all the advantages of free air and
proper moisture, and at the same time
would be secure from the frosts affecting
them, if they were forced early. If they
are thus prepared and planted they would
bear forcing much earlier.

It will be necessary to have a shed at
the back of the house, into which there
should be a door for a passage in very cold

weather,

weather, and to give air in hard frosts when forced early. It would also be very beneficial to have three air-holes in the back-wall above the flue, to carry off the great heat in very hot days when the glasses are opened; and there should be a door at each end of the house for the convenience of passing.

THE frame for supporting the Vines should be at least two feet from the glass, which is much better than when it is nearer; for when they are too close to the glasses, it is impossible, in very hot days, to prevent the sun from burning the leaves and scorching the fruit; besides they have a free air when distant from the glass, ripen much better, and are not so subject to rot.

THE frame should have the same slope with the glasses, that the Vines may be all of an equal distance from it. There should be no glass in front; and indeed there will be little room for any, if the house is built as directed.

THE

THE Vines may be trained ftraight up in front like an efpalier, until they are at a proper diftance from the glafs where the frame is fixed,

IF the Vines are forced very early, it will be proper to cover the border on the outfide of the houfe four or five inches thick with long dung, ftraw, mofs, or any fuch light covering, which will keep the froft from penetrating to the roots, and there will be a free communication with the plants in the houfe and the roots on the outfide.

IT muft be remembered that all Vines fhould be planted in the fame manner that was directed for the firft againft fire-walls, that their roots may run horizontally, and with the fame compound; and that the fide-branches every where fhould be pinched off at a joint above the eye, as before directed.

IF this houfe is planted with plants raifed in heat, they fhould be managed and planted in the fame manner, and at

R 3 the

the fame time, with thofe under glaffes on
fire-walls ; only with this fmall difference,
that in the pruning the firft plant fhould
be cut to one eye ; the next to ten or fifteen, if ftrong ; the next to four eyes';
and fo on to the end ; that is, one, ten,
four.

THE reafon for this difference in pruning is, that here they may be managed fo
as to grow much ftronger than on a common fire-wall ; they have alfo much further to run, fo that the houfe fhould be
furnifhed as foon as poffible, to keep good
wood in all parts of it.

IT would be of great fervice to the Vines
to take the glaffes off the houfe as foon as
the fruit is gone ; but this fhould be done
with care, as they fhould be hardened
gradually, which may be done before the
fruit is all gathered ; for, after the grapes
are all cleared for ripening, there will be
no occafion for much heat, fo that they
may have air night and day.

WHEN the leaves begin to grow yellow

low they fhould be pruned, faftened to
the frame, and the glaffes put on before
the hard frofts and heavy fnows fall; for,
as they are not perpendicular, the froft
and fnow would hurt them ; neverthelefs
they fhould have free air night and day
until the heat is applied.

THE next fpring the fhoot that was cut
to ten eyes fhould have the bottom eye
and the tenth encouraged for next year's
young wood ; and as there will be eight
eyes between them which may have fruit,
they muft be topped two joints above the
clufter. That which was cut to four eyes
fhould have the fourth encouraged for
young wood ; and if the other three have
fruit they muft be topped as before ; but if
any of them, or thofe on the long fhoot,
are barren, pull them off, and ftop the hole
with clay, as before. The fhoot that was
cut to one eye will be very ftrong, and run
to the backfide of the houfe.

THE next pruning feafon the ftrong
fhoot from one eye fhould be cut to the
whole

whole breadth of the house, and it will be full of bearing wood ; the shoot from the top of the plant that was before cut to four eyes should be cut to reach the middle of the house, and the bottom branches to four, five, and six eyes, as they are of strength.

Now the whole frame is covered, there must be young wood left regularly all over at proper distances, as before directed, and at the pruning season shortened as they are of strength.

A HOUSE for Vines built after this manner is much better for forcing early than a common fire-wall, as there is proper conveniency to give warm air in very hard frosts, and in the very dampest weather it may be kept perfectly dry ; for, when there is a long time of dull, damp, rainy weather, it is very difficult to keep the fruit on a fire-wall from moulding without over-heating the wall, which has often been the destruction of both vines and fruit.

In

In the cold winter months, when the forcing is begun, the fires fhould be made very moderately at firft, and have a good deal of air in the day, and a little at night, for fome time; for unlefs the buds pufh ftrong at firft there are little hopes of a good crop.

A week after the fires are made, it will be proper to give a little water to that part of the border which is within the houfe once a week, and it fhould be kept juft moift at all times when the glaffes are on the houfe, although there is no fire; otherwife the roots will decay.

After the fruit is out of bloffom, and about the largenefs of a fmall pin's head, it will add greatly to the fize of the fruit to fprinkle all the floor over in the evening in fine weather, and fhut up the houfe directly, which will caufe a fine moifture like a dew; but care muft be taken to give air early in the morning to dry both fruit and leaves before the fun begins to fhine hot upon them, or it will burn the leaves and fcorch the fruit.   6

IT will be of great advantage to the fruit to give the border a good watering every week after it is out of bloſſom, until it is full grown ; after which they ſhould have none, neither ſhould the floor be any more ſprinkled, but kept quite dry, and have a good deal of air, which will make the fruit of a much better flavour than when they are kept very hot, but they will not be ſo early by ten days. After they are full grown the air ſhould be admitted gradually; for if there be too much given at firſt, ſo as to check the growth of the Vines, it quite ſpoils the flavour of the grapes ; and this is the reaſon why fruit that is forced under glaſs ſhould never have them taken away until all is gathered.

GREAT care ſhould be taken not to draw the Vines weak ; for if they are, there is no other remedy to recover them but a year's reſt ; that is, they muſt be all cut very ſhort, and have no fire put to them before the beginning of May; and then it muſt be moderate, and have a great deal of air.

IT

It will be of great service to sprinkle the floor often to encourage the young shoots; they should also have a great deal of water on the border, both on the outside and inside of the house, as it is not the fruit which is to be regarded this season, but the strengthening of the Vines for another year.

It will be a great disappointment to a gentleman to have few or no grapes for a whole year, who had great plenty the season before; and as every gardener that has the least knowledge of plants must see when they are drawn weak, as soon as that is visible he should give more air and less heat, but this must be cautiously performed; for if, from a great heat and little air, the system was to be changed to the other extreme, it would check the growth of the plants, and not only spoil the next year's wood, but the fruit of this year also.

There are few gentlemen that will be at the expence of having two such houses. If the vines are forced early one year with care, and the next year not until the beginning

beginning of May, as the plants would laſt many years, and bring good crops every year; becauſe at that ſeaſon the weather is fine, and they may have a great quantity of air to ſtrengthen them; for, as I ſaid before, it is impoſſible to force very early without drawing the plants.

VINES in ſuch houſes are ſubject to puſh out many ſhoots from the old wood, which ſhould be conſtantly rubbed off: they alſo puſh roots, which ſhould be taken off as ſoon as they appear.

GRAPES in a hot-houſe that is uſed for Pines in general appear large and fine to the eye, but are moſtly inferior in flavour to thoſe on hot-walls, with or without glaſſes, as they cannot be managed as they ought to be, the Pines requiring a very different culture, and being the principal crop. There ſhould be a good border made in the front, on the outſide of the houſe, of the ſame materials as before directed, which ſhould be four inches lower than the level of the ground, except a foot next

th

the wall where he Vines are planted,
which need not be above an inch. When
they are all planted, cover all the border
with gravel or fand, fo that the walk before the ftove may be neat and clean.

If they are plants raifed in heat, which
are the propereft, plant them four inches
from the front wall of the ftove, in the
fame manner as they are planted in all
other places. Where Vines are intended
in a Pine-ftove, there fhould be a round
hole left againft every rafter when the
houfe is building, and the Vines fhould
be introduced into the houfe as foon as
they are planted, and they will grow a
great length the fame feafon, be very
ftrong, and will bear many grapes the
next year.

There fhould be no more than one
plant to a rafter, which fhould be cut to
a great length when they are pruned ; for
they will be very ftrong, and will produce
fruit at almoft every eye ; but thofe
branches that do not bear fhould be
pulled off.

THERE fhould be a ftrong fhoot trained as near the bottom as poffible every year, and the laft year's bearing branch cut entirely off. There can be no other management of Vines in a Pine-ftove than to keep them clear of all fide-branches, and to top the bearing fhoots at two joints above the fruit; for regard muft be had to the Pines, and not the Vines; fo that fometimes a fmall accident deftroys a crop that had a fine appearance.

As foon as the Vines in a hot-houfe begin to pufh, all the ftem that is on the outfide of the houfe muft be guarded from froft; for one night's hard froft, after the buds are broken, will deftroy all the fruit for that year.

CLEAN mofs, free of grafs and mould, is the beft thing I know for that ufe. It muft be three inches thick all round the ftem, and faftened very clofe.

GRAPES that grow very clofe in the clufter are not fit for hot-houfes, for there is often a great fteam rifes from the bark

when

when in proper order for the Pines, fo that the moifture, in dull weather, is not dried up for feveral days: when the grapes are near ripe, in fuch weather they often rot, if they are clofe in the clufter.

WHEN the grapes grow very clofe, the clufters fhould be thinned when they are very young, and thofe left will grow much larger, and not be fo liable to rot. The large forts that hang loofe are the propereft for Pine-ftoves This work fhould be performed in the morning, when there is an appearance of a fine day, that the wounds may be dried before night by the heat of the fun. This fhould be performed with fharp-pointed fciffors, and great care taken not to wound any of the berries that are left.

THOSE that ufe covers for their Pine-ftoves are in no great danger of having their grapes hurt by frofts; but they are liable to be fcorched by the fun when they are clofe to the rafters. The covering of ftoves is attended with many inconveniencies, fuch as breaking of glafs in high winds,

winds, let the covers be of what fort they will; in wet weather, in treading the fand or gravel into a quagmire; and in very hard frofts, when the covers cannot be taken off for feveral days: in that time, perhaps fome hours of fun every day is entirely loft, alfo the expence of time in covering and uncovering. All thefe things render it very troublefome.

I HAVE not ufed a cover for a Pine-ftove thefe thirty years, although I have had them under my care part of that time, even in the north of Scotland. There is no difficulty in keeping Pine-ftoves with-out covers where there are no grapes; but as they are now introduced into almoft every houfe, it is neceffary to know how to manage them without being at the trouble of covering.

THERE are few ftoves for Pines but where the plants ftand at four or five feet from the glafs. Fix iron pins of a foot long three or four in a rafter, according to the

breadth

breadth of the houfe: there muft be an eye of an inch fquare at the end of each pin, through which place a rod to faften the Vine; it will then be fo far from the glafs as to prevent the frofts hurting the grapes, and they will have a freedom of air, and not be fubject to rot.

THE Vines hanging thus lower will be no detriment to the Pines, although it will fhade them a little; but as it is in the fummer months it will be no difadvantage, as at that feafon there is fun and heat fufficient, although a little fhaded by the leaves of the Vines, and they will be all off before the autumn when fun is wanted. If they have not dropped, they may be pulled off without any injury to the Vine.

GRAPES in Pine-ftoves are for the moft part planted promifcuoufly of early and late forts; the early have in general their berries much clofer in the clufter than the late, and are not fo fit for the hot-houfe.

VOL. I.　　　　S　　　　IF

If the large late kinds are taken into the hot-houfe, they are much higher flavoured than thofe that ripen foon; for as they pufh much later than the early forts, the Pines are put into their proper order for the fummer long before the grapes are of any fize; fo that there will be little or no fteam until the autumn, when the grapes will be moftly ripe.

As the weather in the fummer is warm, the Pine-houfe has a great deal of air, and confequently the grapes will be much richer; and as there is no long continuance of damp weather at that feafon, they will not be fo fubject to rot; there will be this advantage alfo, although they are late in ripening, they will be of no additional expence.

The fire-wall may be forced to bring much earlier fruit than the plants in the Pine-ftove; for the Vines that are in it will not pufh early, unlefs they are taken out of the ftove as foon as the fruit is

gathered

gathered (which is very troublefome, and the Vines are in danger of being hurt) and kept out fome time; as foon as they are put in again they will pufh, but it will be difficult to keep the froft from the roots if they are put in early.

IF the Vines are planted in the infide of the Pine-ftove they will pufh foon and be early; but it is not a good method, for there they can have no water but artificial waterings; and although it is done with judgment and care, it is far inferior to natural rains which fupply them in winter. When in the infide of the ftove they muft be kept damp when they are not growing, or the roots will mould and decay.

IF the fire-walls are properly built they fhould be fifty feet long for Vines to one fire, which fhould be in the middle, draw both ways, and have dampers fo as to throw all the heat one way or both, as fhall be neceffary. One fide may be planted with the early kinds, and the other with a later fort, that will ripen on

S 2

fire-

fire-walls without glaſſes. As they will be every other year without covers, it would be wrong to plant them with any Vines that will not ripen without glaſſes.

THE ſide that has the early Vines may be forced early (two months) before the others, for it will be no detriment to them, as all the heat may be kept from the ſide that is not covered by the damper. It will be much better for the late Vines not to force them before the beginning of March, as then there will be more certainty of a crop, and the fruit will be much better.

To have a regular ſucceſſion for forcing, there ſhould be two lengths of fire-wall in Vines, and one half of each length planted with the early ſorts, and the other half with late.

IF the early ſorts are forced to bear very early, the glaſſes may then be put on the other ſide, and all the heat thrown that way. If it is a cold wet ſeaſon it will be of great ſervice.

THE black, red, white, and brown
Frontiniacs are all very fit for a hot-houfe,
having full as high a flavour as when on
an open wall with fire; the pores of the
berries are much clofer, and do not admit
of fo much of the fteam when the houfe
is fhut up at night as moft of the other
forts of grapes; they never ripen equally
without glafs, and I have had them in
greater perfection in Pine-ftoves than I
could ever bring them to on fire-walls
amongft other Vines. They all grow very
clofe in the clufter, fo fhould be thinned,
as before directed, when very young, which
adds to their fragrancy, as then the fun
gets to all parts of the berries. The red
Frontiniac is the higheft flavoured, and
the moft efteemed of any known in England.

THE Tokay is an exceeding good grape
for the hot-houfe: it is the only proper
place for it, as it is fo long in ripening
after the other forts, that it would caufe
the fire and glaffes to be kept on at leaft a
month

month longer, which would be a detriment to the other plants; and, befides the
expence, they will not come to perfection
without glafs, fo are not fit to be planted
where there is not a conftant cover.

THE brown Hamburgh is a fine large
grape, and grows to a great weight, even
four, five, and fometimes fix pounds a
clufter; but it is not of fo high a flavour
as thofe already mentioned, yet is a good
fruit.

THE Burlake is a fine large grape, and
fit for the Pine-ftove.   It ripens very late.

THE Raifin Grape grows to large clufters;
the berries alfo are very large.   It ripens
the lateft of all the grape kinds, will hang
long on the Vine, and after it is cut may
be kept a month in a dry place where there
is a good deal of air.

THERE are fome other forts that will
anfwer very well for the Pine-ftove; but
                                        thofe

thofe here mentioned are the beft, and will be fufficient to fupply the table a long time after all thofe on the fire-walls are gone.

THE following grapes ripen on fire-walls with or without glafs, but are not fit for forcing early : red, royal, and black Mufcadines ; white, red, and black Frontiniac ; black Sweet-water, black Hamburgh ; all thefe are fit to plant on one fide of the fire-wall, to be forced alternately with or without glafs ; and that year the glaffes are ufed, if you begin about the beginning of March, the grapes will be ripe in July.

THE year the glaffes are not ufed, the middle or end of April is time enough to put fire to them, and then both fides may be worked together. The early forts will be ripe the middle of July, and the late will be all finifhed by the beginning of October.

5

It

IT will be proper to plant the following grapes on fire-walls by themfelves, and they may be forced every other year early, fo as to be ripe the beginning of May; the Sweet-water, brick-coloured Grape, black Clufter, white Mufcadine, and the Miller's Grape; many others will anfwer as well, and perhaps much better, than thofe here mentioned, but I have not had them under my care.

I SHALL not mention any other forts of Vines, as I do not copy from any book, but only relate what has fucceeded under my own immediate direction.

THE laft-mentioned grapes will all ripen on common walls in the fouth of England, and come to perfection.

END of VOL. I.

A
TREATISE
UPON
PLANTING,
GARDENING,
AND THE
MANAGEMENT OF THE HOT-HOUSE.

CONTAINING

I. The Method of planting Foreſt-Trees in gravelly, poor, mountainous, and Heath Lands; and for raiſing the Plants in the Seed-Bed, previous to their being planted.

II. The Method of Pruning Foreſt-Trees, and how to improve Plantations that have been neglected.

III. On the Soils moſt proper for the different Kinds of Foreſt-trees.

IV. The Management of Vines; their Cultivation upon Fire-Walls and in the Hot-Houſe; with a new Method of dreſſing, planting, and preparing the Ground.

V. A new and eaſy Method to propagate Pine Plants, ſo as to gain Half a Year in their Growth; with a ſure Method of deſtroying the Inſect ſo deſtructive to Pines.

VI. The beſt Method to raiſe Muſhrooms without Spawn, by which the Table may be plentifully ſupplied every Day in the Year.

VII. An improved Method of cultivating Aſparagus.

VIII. The beſt Method to cultivate Field Cabbages, Carrots, and Turnips for feeding of Cattle.

IX. A new Method of managing all Kinds of Fruit-Trees, viz. of proper Soils for planting, of pruning and dreſſing them; with a Receipt to prevent Blights, and cure them when blighted.

By JOHN KENNEDY,
GARDENER TO SIR THOMAS GASCOIGNE, BART.

THE SECOND EDITION,
CORRECTED AND GREATLY ENLARGED.

VOL. II.

LONDON:
PRINTED FOR S. HOOPER, Nᵒ 25, LUDGATE·HILL;
and G. ROBINSON, PATERNOSTER-ROW

M DCC LXXVII.

CONTENTS

OF

VOLUME THE SECOND.

CONTENTS.

A

TREATISE

ON

PLANTING

AND

GARDENING.

CHAP. X.

Of the Nursery.

NURSERIES for raiſing foreſt-
trees are very neceſſary for every
gentleman that is fond of planting ; for without a good one no great
progreſs can be made in that moſt delightful and profitable amuſement.

MOST authors who have written on this
ſubject have differed in their opinions, even

so far as to be almoſt quite contradictory to one another. I will not preſume to ſay who are right or who are wrong, but I ſhall give ſome directions that are very different ; and I fix my opinion on this ground, that is, by comparing moſt of the capital nurſeries in England, which are managed by a ſet of men knowing in their profeſſion, and whoſe buſineſs it is to have all kinds of trees in perfection.

It is not intended to give directions for making a nurſery fit to raiſe all the curious trees and plants that are introduced into England, and which thrive very well when taken from the nurſeries where they are raiſed with ſkill and art, and when they are carried into very different ſoils and ſituations all over the country thrive very well ; that would be a taſk too elaborate for the brevity of this treatiſe ; beſides it would be too expenſive for gentlemen, as it is the quantity that makes it worth the ingenious nurſeryman's trouble and care to cultivate ſuch plants.

THE

THE bringing the common forts of foreft-trees to perfection, and making them fit for all kinds of foils and fituations, is what I fhall give directions for, and the choofing a proper foil for that ufe ; or, where it cannot be had, to make one by art, fit for the purpofe, at as little expence as poffible.

IT is a general opinion that all nurferies for raifing trees fhould be the fame, or very near the fame, with the foil the trees are to be planted in ; but this is fetting out on a very wrong principle ; for, as I faid before, all the capital nurferies in England are on a fine light fandy loam, or they are made fo by the nurferymen, who certainly are the beft judges ; and, as a proof of their judgment, all the trees that are taken from fuch nurferies, thrive when planted on much worfe ground than where they were raifed.

I HAVE faid more on this fubject than I firft intended, and am obliged to add a few
words

words more, to prevent a notion that has
long prevailed. I know there are many
gentlemen fo prepoſſeſſed in that opinion,
that they will not allow any other kind of
ground for their nurſeries, than what is
nearly ſimilar to the ſoil they intend to
plant, and this is the reaſon why ſo many
bad-thriving plantations are to be ſeen in
many parts of England.

IF a plantation is to be made on a poor
gravel or a ſtiff clay, what kind of a nurſery would ſuch ground make? all the
plants raiſed on ſuch ground would be
poor, ſmall, hide-bound, ſtarved things,
very unfit for planting in any land, but
more ſo in poor gravel or clay.

To this it may be objected, that I have
given directions for ſowing the ſeeds of
trees on ſuch ground, as the beſt method
for raiſing wood in ſuch ſoils. There is a
great difference between ſowing where the
plant is to remain, and ſowing to raiſe
plants to be tranſplanted: thoſe that are
ſown

fown in fuch ground to remain, twift and
twine their fmall roots amongft the ftones
and gravel, fo as to protect themfelves
from froft in winter and drought in fummer ; but if fuch plants were to be planted with the greateft care, they would be
very liable to fuffer much from the drought
the firft fummer, and be entirely thrown
out of the ground the next winter, as they
would have fo few fhort roots which could
make no refiftance.

It is very wrong to enrich nurferies
with dung. Although the nurferymen
dung their ground very plentifully, they do
it with great judgment, and never plant
trees until it is well rotted, and mixed
with the mould, fo as to be quite incorporated, and generally take a crop of peafe
or beans before they plant ; for if trees
were to be raifed on a bed made rich with
dung, they would grow fo vigoroufly, and
be fo full of juices, that if they were even
planted in very good land, they would be
in danger of being loft for want of a fuffi-

cient

cient quantity of nourifhment, and moft
of them would be, what is called in the
common planting phrafe, hide-bound.

THE proper foil to make a nurfery to
raife foreft-trees, is a light fandy loam of
two feet (if it can be got) or eighteen
inches, which will be a fufficient depth
to prevent drought in fummer affecting
the trees, and the froft hurting them in
the winter.

AN old pafture field that flopes gently
to the fouth is the beft fituation, for low
and flat grounds are not proper, as they
are liable to be over-blown in winter in
deep driving fnows, which will be apt to
break many of the young trees. Befides,
if the fnow be blown very thick, which
often happens, it will lay much longer
than on a rifing ground, and be very detrimental to the young plants.

TRENCHING is always recommended for
making a new nurfery, but it is not always

ways needful; for if the field that it is intended to be made on be clean grafs and free from mofs, plowing will anfwer ; if mofsy, trenching will not be fufficient ; for the mofs, although turned down two feet deep, is long in rotting, and turns to canker, which is very deftructive to young trees.

If the beft fpot that can be found is full of root-weeds, bufhes, or any kind of rubbifh, it will be neceffary to trench, and pick out all the weeds and roots, for there is no poffibility of keeping young trees clean where the ground is full of root-weeds. But it is feldom that needs to happen, as the nurfery may be made at a diftance from the houfe; and if a convenient fpot can be found in a bye corner of the pleafureground, it will not be difagreeable, efpecially to lovers of planting, if it is kept in good order; and its being fo much in view will caufe it to be kept clean.

If the field is a fine clean grafs, and in
A 4 pretty

pretty good condition, plow it up early in the spring, and sow it very thick with the common grey peafe : when they are come into full bloom, plow them all in as deep as the plow can go, and let the ground remain until they are rotten, which will be in six weeks if they are well covered ; then harrow it well with a heavy harrow, and plow it acrofs : in three or four days after harrow it, and before winter plow it again, and let it lay rough all the winter to mellow. In the spring, first harrow, and then plow it ; then harrow it well with a heavy harrow, and it will be in good order for sowing and planting all kinds of forest-trees in the nursery way.

If the field is mossy, pare and burn it as soon as the season will permit : plow it directly, and sow it very thick with turnip or rape seed ; and when it is grown flush, eat it clean off with sheep. As soon as it is eaten bare, plow it as deep as the plow can go ; and when the weeds begin to grow, harrow it first ; then plow, and

3   let

let it lay all the winter. In the fpring, plow and harrow it, and it will be fit for ufe.

If the ground allotted fhould be full of weeds and roots of bufhes, trench it early in the fpring, and pick all the roots as clean as poffible. As foon as it is finifhed fow it with peafe in rows, that the weeds may be kept clean, for in fuch ground there will many annual weeds come up. As for docks, nettles, and quickers they muft be taken up with a dung-fork, for they can- not be deftroyed by hoeing. As foon as the peafe are in bloom plow them in, and when they are quite deftroyed plow the whole again, and harrow it well to get out the weeds, if any remain, and let it lay rough all the winter. Plow it in the fpring, and then it will be in good order.

If a convenient fpot cannot be found that is of a proper temperature, it muft be made fo by art : a ftiff loam, or a light black earth, are the only foils that can be made into a good nurfery. When better
cannot

cannot be had a fandy foil will do ; but it requires a great deal of rich compofition to make the trees flourifh.

A FIELD that has been in corn is very unfit to be made into a nurfery ; but if a grafs field and a ftiff loam, you muft lay over all the grafs a good quantity of fand, two inches thick at leaft ; and if this was done in the beginning of winter, to be wafhed in with the rain and fnow, it would be the beft method.

AFTER it has lain fome time it will cruft over and dry at top, and will be liable to be wafhed off with the rain if the ground has a declivity, which it fhould have ; to prevent that, give it a good harrowing acrofs the field, and that will open the ground fo that the fand will mix much better. Early in the fpring plow it deep, and get from old woods where leaves and fticks have rotted, the bottoms of old wood-ftacks, and the cleaning of ftreets, a large quantity, and lay all over the field

as

as foon as it is plowed ; and when the fwarth has laid long enough to be rotten, plow it acrofs, harrow it well, and plow it again ; then fow it with rape feed, and manage as before directed, and it will be in tolerable good order.

If a light black earth, lay an inch of ftrong loam all over it ; plow it immediately, and then lay fome more loam on and harrow it very well : when it has laid fome time in that ftate, crofs-plow and harrow it, then fow it with rape feed, and manage as before.

By following fome of thefe methods according to the nature of the foil, there will be a good nurfery fit to raife all kinds of common foreft-trees ; only where the feeds are fown, for the planting of poor gravelly land the ground fhould be made lighter than any other part of the nurfery, for the reafons before given.

There are many kinds of foreft-trees,
                                        fuch

such as the Elms of all kinds, that are propagated from layers with great success; and indeed all kinds of forest-trees will grow by layers, so that those who are curious, and find a seminal variety that is remarkably different from the original, the only way to have it preserved genuine is to convert it into a stool, and raise plants by layers.

It has been objected, that forest-trees raised from layers do not grow so vigorous and straight as plants that are raised from seed. There may be some truth in this assertion, but not so much as is generally believed; for it is owing more to the method of laying, than the nature of layers, that occasions this remark. If layers are made from young straight shoots, they will grow as well as seedlings; but when laid from small shoots of a side-branch of an old tree, it is very difficult to make any thing of them but bushes; yet they will take root very well, and this is a good reason why layers are much better than
suckers,

fuckers, for all plants that are to be kept low and bufhy. If a fucker was to be laid it would not alter its nature, it would ftill grow tall and ftraight; but if the young fhoots of an old branch are laid, they will take root and grow, but never fhoot freely : if a fruit-tree, it will bear very plentifully; and if a flowering-plant, it will flower much better than plants that are raifed from fuckers, or plants raifed from cuttings of vigorous young wood.

In the nurfery there fhould be a quarter allotted for the ufe of planting ftools; and as the whole fuccefs depends on the finenefs of the young fhoots that are to be laid, the place where they are to be planted fhould be light and rich ; rich, to encourage the young fhoots to grow ftrong and vigorous; and light, that they may make good roots when laid. The ftools fhould not be planted too thick ; for although they may feem thin when firft planted, if they thrive as they ought to do they will be very large in a few years.

and

and have many layers ; fo that they fhould have room to be laid, and alfo to let the air pafs freely between them : if they are planted in rows, which is beft, they fhould not be nearer than eight feet fquare.

As to the quantity of ground neceffary for a nurfery, that depends on the plantation which is intended. The trees that are to be planted in poor gravel, bare, ftony, and cold poor land, will take up a great deal more of ground in the feedbeds than if they were fown in the common old method ; but as they are to be taken from the feed-bed, and planted out for good, there will be lefs ground neceffary for them than if they were to be tranfplanted from the feed-bed into the nurfery, to remain for four or five years.

It fhould be confidered what fort of ground, and what quantity of each fort is to be planted ; that muft determine the fize of the nurfery. It would be much to

5   the

the advantage of all feedling-trees to have
the beds they are fown in, lay fallow all
the year before they are fown ; and to any
country gentleman that is fond of planting, half an acre of ground more than is
abfolutely neceffary will not be of great
confequence.

THE beds which the feed'ings are raifed
in for poor land muft be quite cleared every
year : they muft be taken up with the
fpade carefully, fo as to break none of their
tap-roots ; if any are broken by accident,
they muft be trimmed, and planted in fome
convenient place of the nurfery, to ftand
for fome time until they are fit for planting in ground that is deep enough to admit of making holes to plant them in, and
they will be as good as any for that ufe,
as all trees that are taken from the feed-
bed to plant in the nurfery fhould have all
their tap-roots cut off.

THERE fhould be no more fown in
thofe beds than can be planted every feafon,
for

for they will not be fit to plant in very
poor ground next year; but if there are
more than are wanted, or can be planted,
they fhould be taken up, have their roots
trimmed, and planted in the nurfery.
They will be much better for that purpofe than thofe that are raifed in beds that
are fown very thick, and ftand fo all the
fummer.

As foon as the beds are cleared they
fhould be thrown into little ridges, and
lay fo till the beginning of May, when
they fhould be dug over and laid flat, and
remain fo all the fummer, for the advantage of keeping them clean; if they
were to remain in ridges, many of the
weeds would be buried in hoeing, and fo
grow again immediately. But it will be
greatly to their advantage to throw them
into fmall ridges before winter, to mellow
them for fowing in the fpring.

THIS management of the ground thrown
into little ridges will be of great fervice,
for

for it will make them produce as fine
and vigorous plants as if there had never
been any thing fown in them; and by
this management they will be in good
order for many years, by only laying
a little of the compoft every other year
upon them, as will be directed to be made
for the recruiting the nurfery-ground,
which muft be greatly impaired by being
conftantly full of young trees.

THE following compofition, if properly
made and laid on, will keep the nurfery
in good order, fo as to produce all kinds
of foreft-trees as ftrong and vigorous, and
free from all blemifhes, as if the ground was
juft taken in, without ufing one grain of
dung, which is a very fcarce commodity
in the country, efpecially where the gentlemen are farmers, and fond of improve-
ments, as moft of them are at this prefent
time.

IF the nurfery ground is ftiffer than
could be wifhed, add fand to the follow-
VOL. II.   B   ing

ing articles; if inclining to fand, which
is the beft, add fome good rich loam, and
it will make fo good a compofition that
its effects will furprize and anfwer the
moft fanguine expectations ; and every
place affords the materials, which may be
procured with little trouble. Collect
docks, nettles, grafs, which fhould be
cut before they come to feed, ftraw, ftubble, rotten wood, leaves, and fhovelings of
the ftable-yard, and make a ridge of them
two feet thick, fix feet broad, and in
length as you can get ftuff, or according
to the largenefs of the nurfery.

IF the ground of the nurfery is light,
put a layer of good loam, four inches thick,
the whole breadth and length of the ridge ;
with another layer of the above things,
two feet ; and then another layer of loam ;
and fo on till there is a ridge eight feet
high : let it lay until winter, and when
the firft deep fnow falls trench it all over,
laying all the fnow in the middle. It may
then lay till the middle of fummer, when

it fhould be turned again, taking great care to keep it clear of weeds between the turnings ; the firft hard froft in winter turn it again, layihg all the frozen parts into the infide, and it will be fit for ufe in the fpring.

SOME time before the feed trees are intended to be fown, lay two inches of the compofition all over the ground, and prick it over two or three times to mix and incorporate it with the old ground ; by being thus worked it will be much moifter (if the feafon is very dry) than land that has been lefs worked : it will alfo be neceffary to prick the beds over juft before they are fown ; and if they are very mellow it will be much better not to rake the beds, but only to level them even with the fpade ; then fow the feeds, and with a flat board prefs the bed level, and cover it according to the fize of the feed.

IT would be a good method to have a large heap of ftuff, the fame as the beds,

B 2   mixed

mixed ready to cover the feeds, which is
much better than taking the earth from
the alleys for that ufe ; for it is trod a
good deal in fowing the feeds, and if the
ground is wet, it is rendered very unfit for
the purpofe ; befides it is of great detriment to the feeds on the fides of the bed
to have deep alleys, as they will be much
drier than if the ground was flufh, and
they will not thrive fo well as thofe in the
middle.

AFTER there is a fufficient quantity of
the above materials collected to make a provifion for three years, it will be proper at
fome diftance to begin a new heap, which
may be gradually increafed as the materials can be got ; but it need not be turned
until there is a quantity fufficient for three
years more, and only keep it clear of weeds
during the time of collecting ; fo that by
having two heaps, one fit for ufe, and one
in gathering, there will be always a provifion for keeping the nurfery in good order.   It will be of great fervice to the
                                         com-

compofition that is ufing, to give it a turning every winter when it is hard frozen, turning all the frozen parts into the middle; but this muft not be done when it is covered with fnow; for although it was abfolutely neceffary to have fnow to mix with the ftraw and other hard dry things to make them ferment and rot at the firft turning, it would be very prejudicial now; it would rob it of its falts, and make it of little value if any fnow was mixed with it.

This compofition is much better for a nurfery than dung, as it will have all the advantages of being kept in good heart, and will caufe none of thofe pernicious misfortunes to young trees that dung is very liable to do: befides, it is attended with no expence except that of collecting it, which is very trifling; and there is this great advantage where there is a compofition of this kind made, that it will keep the grounds in order, as all thofe pernicious weeds, docks and nettles, will be cut and kept from feeding.

THE nurfery muft be put into a regular form ; and as it is abfolutely neceffary to be kept clean, it fhould alfo be neat, and if at a diftance from the houfe it will be an agreeable walk. No gentleman will be at the expence of having a good nurfery who is not fond of trees and planting, and to fuch it is very agreeable to fee a collection of plants of different ages growing, and fome juft rifing out of the grounds which will in time enrich his family, and beautify his eftate and country.

PERHAPS fome people may think too much has been faid about the preparation of the ground, the method of fowing, and the compofition for recruiting the ground after it has been ufed, and that the trees may be purchafed at an eafier expence than by following all the directions that are here given ; and that the chief thing which thofe that raife foreft-trees for planting fhould ftudy, is to have them good, and fit for the different foils they are to be planted in ; but fuch opinions

will

will be found to be wrong grounded, as
the generality of thofe who raife foreft-
trees are at a greater expence than the
work that is here mentioned will coft,
and in general come far fhort of the fuc-
céfs, which will attend thofe that follow
with accuracy what has been directed, befides the advantage of having fine trees of
all kinds to fupply whatever defigns they
intend to carry into execution.

CHAP.

CHAP. XI.

Of Pruning Foreſt-trees.

THERE are a number of people who are againſt the pruning of foreſt-trees, eſpecially the workers in wood, who ſay it occaſions as many blemiſhes as there are branches cut off, and have prepoſſeſſed many gentlemen to be of the ſame opinion. It is not pruning, but the bad methods of performing it that occaſions the objection.

THICK planting in poor land is the beſt method to make them prune themſelves; but that method will not hold good in rich or in middling land, for if the trees were to be planted ſo thick as to prune themſelves, they would draw up weak

and

and be good for little. Befides, trees in good land pufh great fide-branches, which often rob the main ftem and make them grow crooked; whereas in poor land they pufh many fide-branches, but they are weak, and the leading fhoot is generally ftiff and ftrong. To have good and fine timber on rich land, the trees muft be pruned from their firft planting.

Trees planted in good land, or even in middling foil, fhould be at eight feet diftance if they are to ftand for timber; but if the plantation is intended to be confiderable, the trees fhould be planted on good land, at four feet diftance, and they will make a fine nurfery. If the trees are five or fix feet high when planted, in three or four years time they may with great fafety be removed, and another plantation made of them; but if it be at no great diftance, they may be carried with balls of earth fufficient, without being hurt in moving them; fo that there will be the advantage of
having

having another plantation in a year's time as good as the firft. Trees thus planted will thrive better, and be ftronger, than thofe raifed in nurferies, and can be taken up with better roots. It will be attended with a little more expence in making the holes in the field, than if the trees had been planted in the nurfery, but the advantage the trees will receive will more than compenfate for the trifling difference in the expence.

WHEN the trees are taken up to be planted, all the ftrong branches fhould be cut off quite clofe to the bole, and all the fmall ones left at their whole length. The latter end of June, or the beginning of July, the whole plantation fhould be looked over with care, and all the young fhoots that have fprouted, where the branches were cut clofe to the ftem when the tree was planted, muft be pulled off by the hand : they will flip very eafy at that feafon ; but if fuffered to grow much longer, they will be fo hard that the bark

of

of the tree will be in danger of being torn in endeavouring to get them off, and there fhould none of them be cut, for if they are they will the next year pufh out a whole bufh of fprouts at every amputation, and fometimes in the fame autumn, which will occafion double labour, as they muft be pulled off; befides, they grow to a bunch or nob on the bole, and caufe a blemifh in the tree.

In any of the winter months the whole plantation muft be pruned; and all the largeft of thofe fmall branches that were left when planted be cut off quite clofe to the bole; but care muft be taken not to leave the ftem too naked, that is, there fhould be fmall branches left regular all over it. Attention fhould likewife be obferved to leave no branch near the top of the tree equal to the main fhoot : this will prevent its being forky.

In July following, the young fhoots muft be pulled again, and if there is any

great

great diftances on the bole, where there are no branches, a young fhoot may be fuffered to grow to fill up the vacancy, and in the winter, pruned as they were before. This work muft be performed regularly winter and fummer, until the trees have got fufficient length of ftem.

As the trees grow in ftrength, there fhould a foot or two of the bottom of the tree be cleared of all branches every year, and never any more fuffered to grow; but this fhould not be too haftily done, for all trees grow much better, ftouter, and ftronger when the ftem is well furnifhed with fmall branches; for when it is too much divefted of them it grows too tall for its ftrength, becomes top-heavy, and unable to fupport itfelf; is eafily twifted by the wind, and if its not broke, is fo damaged as never to make a good tree, and will be all fhaken when cut. If trees are thus managed for a few years, there will be no blemifhes, they will be very handfome, as well as very valuable, and
when

when cut no fault will be found with their having been pruned.

This method of pruning will anfwer very well for all kinds of foreſt-trees but the Elms. which require a different method to make them fine trees.

The Engliſh Elms are propagated from layers, which if properly performed they will have good roots, and be fit to take up a year after they are laid.   Every kind of Elm will grow very well by layers.

The Engliſh Elm is the fineſt tree of all the kinds of Elm, and in proper foil grows to a very great ſize ; its layers do not take root ſo ſoon as the other kinds. If it was laid in the autumn, as ſoon as the leaves are off, it would greatly encourage its rooting, for the young twigs that are to be laid are hard and dry, and by being in the ground all the winter they are ſoftened, and take root much better. When they are taken up, they ſhould have their

roots

roots dreffed, and planted as foon as poffible, as their roots are fmall and dry faft.
The fide-branches, of which they are very
full, fhould be all cut off at three inches
from the ftem from top to bottom.

IN the beginning of July they fhould
be gone over carefully, and all the fhoots
pulled off, but two of the fmalleft, from
the fide-branches that were fhortened, for
as they will pufh many on every fhortened
branch, they will both rob the main ftem,
and be too heavy for the plant to bear.

IN any of the months next winter the
fide-branches muft be thinned and cut
clofe to the ftem, fo as to ftand regular quite
round the bole, at a foot diftance at the
bottom, and at eight inches towards the
top ; and the fide-branches on the main
ftem of laft fummer muft be fhortened to
three inches.

NEXT July the fprouts that fhoot from
where the branches were cut clofe, fhould
be

be pulled off as before, and the top-branches that were fhortened before muft alfo be thinned.

As the tree grows tall and ftrong, it fhould be cleared at bottom of all branches and kept clean. This fhould be done every year to get a clear ftem ; as the fide-branches, that were left at a foot diftance all round the ftem, begin to grow thick, they fhould be cut clofe off to the ftem ; and if it is not near the bottom fome fmall ones fhould be allowed to grow, and the fide-fhoots that are fhortened upon the main ftem fhould always be thinned the year following.

THIS work fhould be regularly perform-ed winter and fummer until the tree has got fufficient length of bole, and they will be fine ftraight trees, free from all ble-mifhes.

THE common rough-leafed Elm is a good foreft-tree, and for many ufes is pre
ferable

ferable to the Englifh Elm; and it has
another confiderable advantage, that it will
thrive in very indifferent ground; but
it has a great tendency to grow crooked,
to have a large head, with very ftrong fide-
branches.

It may be pruned as the Englifh Elm;
but I know of no method, fo good to keep
them ftraight, as thick planting, and by
that they may be brought to be fine trees.
They fhould be planted firft at three feet
diftance, and in four years, every other
tree, may be taken up and removed into
another plantation, and planted at fix feet
diftance, which is fufficient room for them
to grow to timber.

All the other kinds of Elms may be
managed as the Englifh, for they are all
of them of the fame nature as to their
culture, but far inferior as to their utility,
efpecially the Dutch, which thrives very
well for twenty years, and then in general
is at a ftand; befides, the wood is of no
great value.   2

THE Elms of all kinds, although they have never been pruned, and are grown very rude, may be reduced to order without any detriment to the wood; but the whole bole will puſh young ſhoots after they are pruned, as well as where there are branches cut off, and will take a good deal of labour to keep them clean for three or four years.

IF the trees are ſtrong and ſtiff in the bole, for they may be pruned at any age, the head ſhould be left round and handſome, which greatly hinder the ſhoots from growing ſo numerous on the bole, and they will ſooner give over growing; but if the bole is ſlender, the top ſhould be lightened to the very laſt year's ſhoot, but none of thoſe towards the top ſhould be cut cloſe; and it will be ne-ceſſary to leave ſome branches two or three feet long, but they ſhould be as equal round the bole as poſſible.

THE pruning of Firs and Pines may be

performed with tolerable safety; but it is not to be recommended as a good practice. The best method is to plant them in clumps, even on good land, at ten feet distance tree from tree, and they will prune themselves.

But if Silver or Spruce Firs are intended for ornament, they should have fifty feet at least, and they will make a fine shew; but as the ground round would look naked for many years, the number intended to stand should be planted first, and then the spaces between may be filled, so that they may stand at ten feet distance; but they should be removed before the branches meet, for if they are destroyed it spoils their beauty.

If there is a necessity to prune Firs or Pines that hang over and spoil better trees, it should be done in the winter, and no branches be cut nearer than two feet to the bole; for if they are cut close they make a blemish, but if cut long they decay gradually and do no hurt.

The method of pruning here directed may feem to take a great deal of labour, and be a very great expence, more than the advantage the trees will repay. By being pruned the trees will be of a much greater value, and the expence will not be much, if regularly performed every year. As to the pulling the young fhoots at the proper feafon, for then they will come off very eafy, a man will go through a great many trees in a day, even if they are large; for although the directions are long, they could not be abridged to be made plain, as they are not in general practice; but when once they are regularly carried on, they will not be expenfive, and will anfwer beyond the moft fanguine expectation.

When the trees are fmall and can be eafily bent, they may be pruned and pulled by the hand, ftanding on the ground; but when the trees are become ftrong and ftiff, it is not a good way to bend them, for although they may feem to receive no hurt, it certainly ftrains the bark on both fides, and muft be a great detriment.

WHEN the boles are so weak that they
cannot bear a ladder, and are too stiff to
bend, the best way is to have a light pair
of steps six or eight feet high, which will
be a sufficient length to clear the bole until
it is strong enough to bear a common ladder; and the common pruning irons will
dress the tops: there should be three
sizes of them; they are called half-moons.

THE management and right ordering
plantations on good, or even middling land,
is of the utmost consequence; and unless
they are properly taken care of will never
be very fine timber. Trees on good land
will thrive, and there will be many fine
ones amongst them, although there is no
care taken of them after planting; but
there will be many bad ones to one good
tree. If they had been dressed, all or
the greatest part of them would have been
good timber.

THERE are to be seen in natural woods,
and also in plantations, many mere bushes
                                    with

with boles not a yard long, which would have been good trees, had there been a little labour beftowed upon them. When the following directions are put in practice, they will remedy all thofe evils, and fave many a tree.

On poor land, as has been before related, thick planting will anfwer the ends of pruning and dreffing; but on good ground it will not anfwer to make good trees, unlefs proper care be taken of them; and they muft be treated in quite a different method from thofe on poor ground; for as the trees grow much fafter, they will require care and labour to keep them in order, but no great expence if regularly performed.

If trees on good land are planted thick, which is the beft method, they muft be pruned regularly fummer and winter, as directed, for they muft never be allowed to grow to be thickets, for that would draw and fpoil them entirely: and as they

C 3   are

are planted thick, to remove when they
want room, to encourage their growth at firſt
planting, and all the trees being regularly
pruned, when they are to be removed
you have only to dreſs their roots, and
move them into another plantation.

IF they are to carry only a little
way, thoſe taken up in the morning can
be planted in the afternoon, their roots
will require but little dreſſing (if properly taken up) although there be no
earth carried with them; but if they were
moved with ſmall flat balls, which may
be eaſily done, as they will have good
flat roots, and there would be no occaſion to cut any more than the ſmall ſtrag-
gling roots that were cut with the ſpade
in taking them up.

THE expence of carrying them with
balls will not be great, and gentlemen that
are fond of planting would not regard
ſuch a trifle to have another plantation as
large as the firſt, which in two years it
will be impoſſible to diſtinguiſh which

is the beſt, or the firſt planted.   All
plantations on good land ſhould be planted
thick, and managed in this manner, and
if they were would be fine trees.   But
there are many plantations on good land,
and planted thick, that are never pruned,
nor a tree removed, which are now mere
thickets of ſmall and unſightly trees.

C H A P. XII.

Of *Pruning Fruit-Trees.*

IN my treatiſe on gardening, I gave par-
cular directions for making Kitchen-
gardens in ſuch a manner, that neither
the roots of the wall-trees, nor thoſe
planted round the quarters, for dwarfs or
eſpaliers, can meet with any obſtructions
to canker and impede their growth.   The
method in practice, is to mark out the
walks in kitchen-gardens, and to remove
all the good earth, and make them the common receptacles for all manner of rubbiſh
during the making of the whole garden.

THIS method is very prejudicial to the trees round the quarters, which seldom have above four feet to spread their roots, before they strike into the rubbish. This is one cause of their being short-liv'd, and producing so indifferent fruit after the fourth or fifth year.

IT is a general complaint that all directions for pruning are so prolix, that they are difficult to be understood, even by those of experience in that branch. To remedy these defects I shall use my utmost endeavours to render this treatise as plain and easy as possible.

THERE are many things materially necessary to be considered besides pruning, in order to have healthful trees, without which it is impossible to have good flavoured fruit.

THERE are gardens that produce fine fruit of a high flavour, and others contiguous to them, whose fruit are not much superior to crabs in relish, although managed

managed (as to pruning) both exactly
after the fame method.

THIS difference of flavour is generally
attributed to the fituation, but never to
the foil, from a fuppofition that there is
little difference.

A GOOD fituation is certainly of great
advantage to the flavour of fruit ; but if
the foil is improper, though ever fo fkilfully pruned, the fruit will not be high
flavoured, although often fair, and beautiful to the eye.

THE general method in making the
borders in kitchen-gardens, is to make
them rich, without regard to the different
kinds of fruit to be planted in them.

WHEN thus prepared the trees will often thrive, look well, and produce great
quantities of fruit ; but it will be very inferior in flavour to the fruit of thofe
trees that are planted in foils which are
properly adapted for them.

To remedy this evil, I fhall give directions for making the borders fit for all kinds of fruit ; and to render this eafy, it will be proper to plant a good many of the fame fort together, and not promifcuoufly, which is the general practice ; for if they are promifcuoufly planted, it will be difficult to prepare the borders properly ; in that cafe there muft be a change in the preparation every five or fix yards.

In fmall gardens, where there is only room for a few trees, perhaps one or two of a kind, it will be eafy to prepare the borders at firft ; but difficult to keep fo many parcels of compoft for recruiting them : when that happens, to prevent trouble, all the borders may be made of two forts, viz. for peaches and cherries (as will be directed under that head) which will anfwer tolerably well, and is the beft way where a gardener is not kept.

In large gardens, where there is a large collection of fruit-trees, the preparation

anfwers

anfwers for the fame kind of fruit on all afpects. There muft always be a quantity of each fort of the compoft ready prepared, to lay on the borders every third or fourth year; this will keep the trees many years in good heart, and the fruit will be very high flavoured.

It will be of great advantage to the compoft for the borders to be turned over three or four times in a year, and to be two years old before it is ufed.

It will add greatly to its fertility to be turned over in winter when its hard frozen, and all the frozen parts turned into the infide; it fhould not be turned when covered with fnow, unlefs it is fwept clean.

Having now given directions for the management of the compoft, before I proceed any further, fhall give general directions for planting all kinds of fruit-trees, on walls, dwarfs, and efpaliers, on borders and in orchards.

Wall-

WALL-TREES fhould never be planted nearer the wall at bottom than nine inches; if they are planted fo clofe, for the bole to prefs againft the wall, it often gums and cankers them.

THE proper diftances being marked on the wall (which fhould be afcertained for each kind of fruit) open a hole a foot fquare, and fixteen inches deep, in which lay a flat ftone at leaft two inches thick, and on the ftone three inches of the mould prepared for planting.

THE roots of the tree muft be pruned fo to ftand floping on the mould laid on the ftone, the head inclining to the wall; fill up the hole with the planting mould, and tread it gently; then loofen it an inch deep with the fpade, after treading it. Cover eighteen inches round the bell of the tree with mofs, two inches thick, preffing it flat with the hand; faften the tree to the wall with a fingle nail and fhread, to prevent its being fhaken with the wind, but fo loofe that the tree

may

may not hang by the fhread, if the ground
fhould chance to fink. There is nothing
further neceffary, until they are headed
down in fpring, which fhould never be
done before the buds begin to fwell.

THE trees for walls are generally
brought from fome diftant nurfery, and
are fome days out of the ground; for
which reafon all the fmall roots muft be
cut clofe off the main roots, and thefe
properly fhortened, and never left croffing
one another.

IT is preferable for all kinds of wall-
trees to have but one ftem to be young
and vigorous. If the ground is tolerably
dry in autumn (after the leaves are fallen)
the fooner fruit-trees are planted the better; but if the bottom is of a cold watery
nature, the fpring is preferable.

IF the trees to be planted are on the fpot,
they may be removed with fafety, although
the leaves are frefh: when that is the cafe,

there

there is no occafion to cut off any of the
fmall roots; but if they are out of the
ground a day, the fmall roots dry, the
bark fhrivels, and the tree often decays;
for which reafon, it is beft to let the leaves
drop before the trees are removed to a diftance.

CLEAN mofs is preferable to all kinds
of ftraw or dung, to lay round new planted
trees; it breeds no vermin, and keeps out
the froft and drought: great care muft
be taken not to bring grubs with it from
the field, for it is often pulled up in great
pieces, in which there are many: it
would be worth the labour to leafe it all
over before it is laid round the trees.

THE planting of dwarfs or efpaliers in
the borders of kitchen-gardens, the making of the holes, and laying the flat
ftones, is the fame as for wall-trees; but
there is fome difference in the manner of
pruning the roots.

IN

IN pruning the roots of wall-trees, those on the side next the wall should be all cut off, as there is no occasion for any but those that point from the wall; but for dwarfs or espaliers, the case is quite different; the roots should be cut to spread as regular as possible all round, that they may be able to defend the tree, let the wind blow from what quarter it will.

IF the trees for dwarfs or espaliers are planted in autumn, it would be best to fasten them to a stake, and head them down in spring. A wet bottom is very bad for a kitchen-garden, especially to all kind of trees; for although some of them may thrive tolerably well, and bear a great quantity of fruit, it is never good, if the soil is prepared with all the art imaginable.

ESPALIERS are now banished all good gardens, for many reasons: if the trees are on paradise-stocks, they are of short duration, and often decaying in patches, which makes them very unsightly. If on stocks,

ftocks, unlefs they are allowed a great deal
of room to fpread and to grow to fix feet
high, they require fo much cutting to
keep them in order, that they feldom produce much fruit.

WHEN the efpaliers are allowed to grow
high, unlefs the quarters of the kitchen-
garden are large, in fummer they caufe
the herbage to draw up weak, it is then
never fo good, nor fo well tafted as when
it has free air.

ESPALIERS are alfo ftiff and formal,
and fpoil that agreeable rural look of trees
growing in the natural way.

APPLES on French paradife-ftocks,
planted at eight or nine feet diftance,
pruned and kept in an eafy manner, make
a fine appearance, and produce better fruit,
and in greater quantities, than when they
are in efpaliers.

TREES planted and trained thus, admit
free

free air into the quarters; and the little openings give a view into them which is pleafant to thofe that delight to walk in a kitchen-garden.

If the kitchen-garden is large, the trees on the fouth-fide of the quarters, behind the north wall, the infide of the garden, may be planted with apples on Dutch pa- radife-ftocks, and allowed to grow as high as the wall; it will be very agreeable in fummer: they laft much longer than the French-ftocks, and will bear more and finer fruit; the French paradife-ftocks are apt to canker, but if the walks are good earth, the fame as the borders, they will not be fo liable to that misfortune.

To prevent any unfightly trees in the borders round the quarters, it would be right to have a few fpare ones growing in a corner of the garden, which might be taken up with a bole, and put into the place of any tree that is cankered or decay- ing. If the tree that is taken up is not

VOL. II.   D   far

far gone, it may be planted in some bye place ; moving often ſtops the canker. The only objection to training dwarfs in this manner is, that the fruit is more liable to be blown off than from eſpaliers ; if the firſt ſhoots are trained horizontal for two years, they will grow ſtiff, and will not be much hurt by the ſtrongeſt winds. This manner of training will alſo anſwer for common orchards, the fruit will be much eaſier to gather, and not be ſo ſubject to be blown off by the wind, as when the trees are high : it may be objected, that the cattle would crop the lower branches.

The laying flags at the bottom of all fruit-trees is good in all kinds of ſoil. If the roots are dreſſed, and the trees planted as directed, they will never go lower, but ſpread horizontally ; they will continue many years, and bear excellent fruit.

It is recommended by ſome, and the practice of many, to lay a quantity of rubbiſh

bifh in the bottom of borders, to prevent the roots getting down into clay, fand, or gravel: this never anfwers; the roots will ftrike into the rubbifh, and even through it, if a foot thick, into the fand, &c. but as foon as the roots reach the rubbifh, the tree cankers and the fruit fpots.

The roots of fruit-trees fhould not be above one foot deep in the ground, for the foil below that is hard, dry, and full of rancid vapours, even in good foil. The nourifhment the roots draw from thence fpoils the rich flavour that thofe fruits have whofe roots are no deeper than the air and rains penetrate.

It is the general opinion that old trees cannot bear good fruit on account of their age; this is feldom the cafe; the reafon indeed is, all the fmall roots are fpread too deep into fand, gravel, or clay; hence they canker, and the fruit is fpotted.

D 2   Apples

APPLES on crab-ftocks will laft many years, and bear good fruit. An inftance of this I faw in the ruins of a monaftery which had been in the fame family ever fince its diffolution, and by tradition the fame trees that were in the place when it firft came into their hands fome hundred years ago.

THE trees were much decayed, but what were alive of them bore fair round fruit, equal except as to fize to any tree of ten years old. The whole orchard was paved with bricks; the foil twenty inches deep, a fine rich loam. There was a plantation of pear-trees about thirty years old; which had covered the wall fome years, and produced great quantities of fine fruit; at laft they began to canker, and the fruit to pit: they became every year worfe. But the following experiment brought them to flourifh again:

THE ground was opened all round the bole of the tree at three feet diftance; the

roots

roots were cut off all round at that diftance ; the bole thinned to the thicknefs
of two feet, a ftone put under it, and the
whole filled up with a good frefh loam.

THIS was performed in winter ; it was
late in fpring before they came into leaf.
They made no fhoots, and the few fmall
leaves they had foon decayed : they had
fome water in the fummer. This is a
proof of the great utility of preventing the
roots from ftriking too deep into the
ground.

NEXT fpring they were in leaf as foon
as any of the fame kind, made little wood,
but clean, and had fome fruit, which
was fair and clean. The third year they
were as vigorous as when firft planted,
quite clear of canker, and produced a
great quantity of fine fruit.

HAVING given directions for planting,
with fome reafons for what has been faid
on that head, we fhall now proceed to

prepare

prepare the borders for each kind of fruit, adapting the foils that will preferve the trees healthy, and bring the fruit to its utmoft perfection in fize and flavour.

THE foil and fituation fhould be confidered in fixing on a fpot for a kitchen-garden, for if there is not a foot or more of good foil the expence will be immenfe, if the garden is only of a moderate fize. The beft natural foil for a garden is a light loam, and where eighteen inches deep of fuch a foil can be got the expence will be trifling.

To prepare Soil for Apricots.

THE favourite foil of the apricot is a light loam : if the natural foil is a rich loam of eighteen inches deep, dig from a common as much light fandy earth as will fpread fix inches thick all over the border ; to every load of fandy earth add one barrow of rotten dung. If the natural earth is fandy, add one third of rich

loam ;

loam ; and to every four loads of loam add one of rotten dung. If the natural foil is gravelly, add one half ftrong loam, and to every three loads of loam one of rotten dung, and one of rotten wood earth, if it can be got. The compofition fhould be laid on the border and trenched over three or four times that it may be well mixed ; the laft time fhould be three weeks before the planting feafon, that the mould may be well fettled before the trees are planted.

THE apricot is naturally inclined to fhoot ftrong vigorous wood, efpecially when the border is rich ; dung is perni- cious to all trees (vines excepted) but none fuffer fo much from it as apricots, for it makes them gum and canker.

APRICOTS fhould have more room than is generally allowed them ; the Turkey kind fhould never have lefs than thirty feet, al- though the wall is twelve feet high, and the other forts from twenty to twenty- four feet.   D 4

When the planting feafon is come, mark out the diftances, open the holes, and difperfe the mould all over the border. Some time before planting prepare the following compofition for that purpofe, viz. four barrows of earth from that prepared for the border, one barrow of very rotten dung, and one barrow of light rich black earth ; this is the proportion, the quantity muft be according to the plantation ; it muft be well mixed and a barrow and a half laid to every hole. The little quantity of dung ufed in planting can have no bad effect, it will make the treo pufh forth fine ftrong roots the firft year.

*To prepare Borders for Peaches and Necta-
rines.*

Peaches and nectarines are fo much alike in nature, that they thrive very well on the fame foil ; what is faid of one may be underftood of both : the right preparation of the borders is very material. as on this the future fuccefs depends.

Peaches

PEACHES love a ſtrong loam, in which they thrive beſt, and will come to great perfection. Dung is a great enemy to them, as it cauſes them to ſhoot ſtrong rambling wood, which is very detrimental to the trees ; it alſo cauſes them to gum, and prevents their bearing.

IF the natural ſoil is a ſtrong loam, add one inch of very rotten dung, and trench the border over three times.

IF the natural ſoil is gravelly, add one half of ſtrong loam, and two inches of rotten dung ; if ſandy, which is the worſt of all ſoils for peaches, add three inches of ſtrong loam to one of the natural earth and one inch of rotten dung ; if a fine light rich earth, add one third of a good ſtrong loam. The borders ſhould always be trenched over three times, after the proper mixtures are laid on before planting. The following compoſt muſt be prepared for planting : four barrows out of the prepared border, one of light black earth, and
one

one of very rotten dung; one barrow to each hole. The proper diftance for peaches and nectarines is feldom properly confidered; they are in general planted at equal diftances, without regard to their growth, though many forts require a great deal more room than others.

The early forts of peaches and nectarines fhould be planted from fourteen to fixteen feet, the late fort from eighteen to twenty feet. It is a general miftake in planting walls with peaches and nectarines to have a great number of forts, that there may be a variety all the feafon; fix or feven good kinds properly chofen are fufficient to afford plenty and variety during the feafon.

There are many peaches fo much alike that it is difficult to diftinguifh them even by good judges of fruit.

It is a common practice to plant fire walls with thofe that feldom come to per-fection in England without heat.

By this method, it is true, fine fruit may be obtained that cannot be had otherwife, but where there is not a great quantity of walling it is attended with many inconveniencies.

It is abfolutely neceffary to allow the trees reft, at leaft every third year ; the late forts will not then ripen their fruit or wood if the feafon is not favourable ; when that is the cafe, much of the young wood will be hurt by the froft in winter, and the tree fo mangled, that there is often a difappointment upon an increafe of the fucceeding year.

If the walls are all planted with good kinds that are eatable in favourable feafons without heat, by the help of fire they may be brought to the greateft perfection, and in the year that there is no fire, they ftand a good chance of ripening their wood if the feafon is but indifferent.

The

The Preparation of Borders for Pears.

THE propereſt ſoil for all kind of pears
is a ſtrong loam : when the natural ſoil
is ſuch, add one inch of very rotten dung,
and trench the border over three or four
times.

SAND and gravel are great enemies to
all kind of pears ; on ſuch ſoil they moſs
and canker, and never produce good fruit ;
it is generally ſtony and has no flavour :
where the natural ſoil is ſuch, there muſt
be added a great deal of loam, and two
inches of very rotten cow-dung.

IF the natural ſoil is a ſtiff clay, in that
caſe it will be proper to raiſe the border
eight inches above the level of the ground,
which muſt be with the following materials : coal aſhes ſifted very fine, wood
earth where leaves and ſticks have rotted,
ſoft ſand from a pit, and rotten horſe-dung
of each an equal quantity, to be laid on
the

the clay, and worked over until it is well mixed; in which pears will thrive and produce moſt excellent fruit. The mould for planting in this ſtiff border is, one barrow from the border, one of rotten dung, and two of wood earth : this will be fit for the trees to ſtrike root in, after which they will grow very well.

No kind of pears ſhould have leſs room than twenty feet, and many of the more vigorous ſorts ſhould have twenty-four or thirty feet diſtance at leaſt ; for when they have little room they grow ſo full of young wood and require ſo much cutting, that they never bear well.

THE compoſt for planting with (except the clay border) is, two barrows of the prepared mould, one of rotten horſe-dung, and one of light rich mould ; put one barrow of it to a hole.

PEARS are a fine fruit and laſting, if brought to perfection. There are many of

3   the

the French kinds that are equal in goodnefs, if not fuperior, to many peaches:
an agreeable entertainment in the winter
months.

THOSE that fooneft come to perfection in
England are, the winter Boncrêtien, the
Chaumontelle, Eafter Bergamot, Virgoulé,
Colmar, Beurré, Crafan, and St. Germain.

THERE are fome of the above, viz. the
Crafan, Beurré, St. Germain, and fome
times the Colmar, prove tolerably good in
fine feafons, but are fo inferior to the fame
kinds in France, that if compared together they would appear a different fort of
fruit. By planting them on a fire wall
and giving them a little heat from fetting
until they are fit to pull, it improves them
beyond the conception of thofe who have
not feen the experiment. In fmall gardens
where the fouth walls cannot be fpared, if
the weft afpect is flew'd it will anfwer for
pears and is much better for apricots, the
fruit is larger and much better flavoured;

5   cherries

cherries likewife are larger and not fo fubject to vermin as on a fouth afpect.

The Preparation of the Borders for Plums.

GRAVELLY light foils are the beft for
all kinds of plums ; they bear high flavoured fruit in great quantities ; they are
not fo large as when planted in ftrong
earth, but the quantity and richnefs of the
flavour make amends for that deficiency

IF the natural foil is a loam, add an
equal quantity of poor light fandy mould
from a common.

IF a rich black foil, add one third of
fandy loam and one third of poor gravelly
earth from a barren common. If the natural foil of the borders is a light fand,
add one half of a ftronger loam : if the
natural foil inclines to clay, it is very unfit for plums ; it muft be made light or
there will be little hopes of fuccefs. One
half of the earth muft be taken out of the
border

border and replaced with light rotten wood earth if it can be got, or with rich black earth: to every load of earth add one of fine small gravel. To all the different soils add one inch of very rotten horse-dung, then trench the whole border over three or four times; the oftener the clay border is worked the better. In all of these preparations plum-trees will thrive and bring their fruit to great perfection.

THE proper diftances for plum-trees are from fixteen feet to twenty; it is much better to have fewer trees and allow them room to fpread; they will be the handfomer, and bear a greater quantity and better fruit; for trees that are crowded produce much wood and little fruit.

To prepare the Borders for Cherries.

CHERRIES thrive eft in a fine light rich loam; in fuch they bear great quantities, and the fruit has a high flavour. If the

the natural foil be a ftrong loam, add a
large quantity of foft pit fand until it is
almoft a fandy loam : if a fandy foil they
will thrive tolerably well, but if three or
four inches of ftrong loam were added, the
trees would be more vigorous and the
fruit much larger; in a fandy foil they
will be fooner ripe by ten days than in any
other mould : if the natural foil is a
light black earth it will anfwer very
well. If the natural foil is a ftrong loam
inclining to clay, add foft fand, rotten
wood earth or any other light foil to
make it light.

To all of thofe different mixtures muft
be added two inches of very rotten dung,
that has been turned feveral times. The
whole muft be trenched over three or four
times that the border may be well mixed
before planting. The diftance for cherries
of all kinds is from eighteen feet to twenty-four ; this may feem a great diftance,
but there will be more fruit on one tree
that covers twenty-four feet of wall, than

there would be on two trees standing on
the same length of ground.

The only objection that can be made to
this great distance is, that it will be some
years before the wall is covered; but if
the method of training the trees here directed be followed, they will soon cover it : however, to remove that objection, plant standards between, which
may run into fruit without any regard
to the trees, for they must be cut out as
the dwarfs advance.

The composition for planting is, four
barrows of the prepared mould for the
border, to one of very rotten horse-dung.

In the composition of the mould for
planting, there should be always ready
some of the prepared border mould ; what
is meant by that is, some of the mould of
the border, the trees are to be planted in,
after it has been trenched and well mixed:
this is to be observed in all the different
kinds of soils.

To prepare the Borders for Figs.

FIGS thrive only in a fine light rich earth; their large roots are long and smooth, and pufh out many fmall fibrous roots which are too tender to make their way into a ftiff mould; on this account there are little hopes of fuccefs, unlefs they are planted in light rich mould.

IF the ground is gravelly or fandy, the cleaning of a pond that has not been drained for fome years, and rotten wood earth might be added, until there is fuffi-cient to make it light and rich.

THE diftance a fig-tree will fpread on a wall in proper foil is very great: there were fome old fig-trees of a large fize which covered a great length of wall in many parts of England that were greatly hurt in the year 1739.

FIGS fhould be always planted at the diftance of twenty feet from each other;

for if there is not room to lay in young wood, there will be very little fruit. The composition for planting thefe trees is two barrows of fine light mould, one of rotten horfe-dung, and one of rotten wood earth. It fhould be obferved that in the directions for planting mould, the quantities muft be proportioned to the fize of the plantation.

THE proper foil for an orchard is a fine light loam, they will grow and bear fruit in all good earth that is twenty inches deep, and a dry bottom.

CLAY, fand, or gravel are very unfit to plant fruit-trees in; for though the ground be well prepared before planting, they foon decay.

WALNUTS, if planted for fruit, fhould have a good, light, rich, deep foil. The trees raifed in the nurfery, that have been removed at leaft three times, are the propereft; the top-root being deftroyed, the

5　　　　fide-

fide-roots run horizontal, and then they bear great quantities of fine fruit.

CHESNUTS planted for fruit, fhould be treated in the fame manner as walnuts; but they will thrive in worfe ground.

MULBERRIES fhould be planted on a dry light earth, not too rich; the ground all round them for fix yards fhould be covered with grafs; for if it is dug, they never bear any quantity of fruit, and what they do, will be very indifferent.

FILBERTS will thrive, and bear great quantities of fruit, if planted on a dry, light gravelly foil, and the fruit will be much fweeter, than the fruit of thofe planted in rich or ftrong land. If the walks in the kitchen-garden are of the fame foil with the quarters, there is no neceffity for the borders to be very broad; but if they are lefs than ten feet, the walks fhould be prepared the fame as the borders, before the gravel or fand is laid on, which fhould not exceed four inches.

IF there is a fruit wall near a manſion, it would be neater to have the gravel laid cloſe to the wall: it will be no detriment to the trees, provided the ground is properly prepared before the gravel is laid on; and once in three years it muſt be taken up to lay ſome freſh compoſt to the trees.

THE preparing the fruit borders of different kinds, ſo that each ſort of fruit may have its proper ſoil, is not ſo great an expence as what is beſtowed in the common method of making all the borders in the garden equally good.

To keep the trees in heart, and the fruit in perfection, there muſt be ſome freſh compoſt laid on the borders every third year.

IF theſe directions are carefully obſerved in preparing the borders, and planting the trees, there will be no doubt of having good trees and fine fruit, provided they can be kept from blighting

THERE

THERE have been many things pre-
scribed to prevent and cure blights, none
of which have yet been found effectual.
However, there are a variety of things
that are great helps; and it is my opinion,
the reason of their miscarriage is owing
to the directions not being duly observed.

I SHALL give a receipt that has done
great things; and where the directions
have been minutely followed, have never
failed, as I could hear of.

A Preparation to prevent Blights.

PROVIDE two tubs that will hold two
hogsheads each, if the garden is large and
a great number of trees; if a small garden,
tubs that hold one hogshead each will be
sufficient.

PUT into one of the tubs two pecks of
clote lime; fill up the tub with clear water, stirring it up from the bottom.

E 4 NEXT

Next day draw off the water, as long as clear, into the empty tub, fill up the lime tub with clear water, stir it up, and when clear, draw off as before. This must be repeated every day until there is a hogshead of clear lime water.

To a hogshead of clear water must be added six pounds of flour of brimstone, and four pounds of tobacco dust, which is difficult to mix the sulphur and dust with water : take a small tub, into which put the sulphur and dust, add a little water, and mix them gradually, adding more as it grows wet ; thus you must proceed until the whole is mixed : it must then be put into the tub of clear lime water and well stirred ; it is then fit for use.

There is a liquid that is squeezed from the tobacco in pressing, which is much better than the dust ; those that are near a tobacconist, may get it for a trifle. One pint of it instead of the four pounds of dust. A hand-engine is very proper for

washing

washing the trees, but where there is not one, it may be done with a water-pot and rose, standing on a ladder.

A PECK of fresh lime added to that first put in, will make six hogsheads of clear water, which must be drawn off as before; the same quantity of sulphur and dust (or liquid) must be added to every hogshead of clear water. The trees should be washed as soon as the buds begin to burst, at least three times a week. At that season the nights are in general frosty, therefore the trees should be washed between seven and nine in the morning. When the season is farther advanced, it will be found necessary sometimes to wash them till the beginning of June; the frosty nights being then over, they may be washed from five to seven o'clock in the afternoon.

WHEN the leaves begin to spread, if any of them curl, they should be pulled off. This wash is also good for goosberries and

and currants; it has brought peach-trees
to flourish that were thought past all recovery.

THE soil being improper often causes
them to blight, and to grow in such a
rude manner that the best instruments in
pruning cannot keep them in order; when
that is the case, the soil and depth of the
trees roots should be examined. If the soil
does not correspond to any of those kinds
directed, for the kind of fruit growing *in it*,
the border should be properly prepared,
and if the roots have got too deep, they
should be raised.

IF the roots of peaches or nectarines are
too deep, and the trees above eight or nine
years old, it is better to plant near one;
all other kinds of fruit may be moved after bearing thirty or forty years.

HAVING now gone through the preparation of the borders, planting the trees,
and given some directions to prevent their
blighting,

blighting, I now proceed to the management and pruning of them.

The Management and pruning of Apricot Trees.

THE spring after planting apricot-trees, as soon as the buds begin to push, head them down, if the trees are healthy and strong, to six eyes, if weak, four will be enough.

RUB off the fore-right shoots that are produced on the stock, and nail the side branches as soon as they will reach the wall.

IF the tree was left with six eyes, there will be at least two shoots of a side; if to four eyes, two branches of a side, which should be nailed horizontally at five or six inches distance. The latter end of October they should be pruned, if they have made vigorous shoots, to eight, nine, and ten inches; if weak, three, four, and five, will do: they should be nailed directly.

In the spring, when the buds begin to push, all the fore-rigne eyes should be rubbed off, and the young shoots laid regularly in from the last year's wood, at five, six, and seven inches distance, which should be nailed as they advance in length all the summer. It is the common method to spur apricots; but it is better to keep them full of young wood; the fruit is much larger, and the blossom is not so liable to be killed by the frost in spring, as that on the spurs which is so far off the wall.

The next October the young shoots must be shortened according to their strength, to four, six, eight, and ten inches: perhaps there may be some very vigorous shoots, which must be cut to eighteen inches or two feet long, if there is room to lay it in and the young shoots that come from it, if not cut it clean off

In spring the fore-right buds must be rubbed off: as the tree is now large, this work must be performed at different times,

and

and the wood for next year laid in from time to time until the tree is well furnifhed all over.   The young wood fhould be laid in from the buds that ftand fair, on the fides of the laft year's fhoots; and none fuffered to grow but thofe that are laid in for wood.   It muft be obferved, that no ftone fruit is fond of being cut at this fea-fon : much work may be performed in a little time by rubbing off all fuperfluous buds.   If this method is followed the tree will be handfome and produce good fruit, and will not be fubject to gum, which occafions the lofs of many a tree.

WHEN a ftrong luxurious branch is produced in any part of the tree, it is beft to cut it clofe off; for if it is fhortened to produce wood, which is often recommended, it never anfwers, for the fhoots that come from it are never good, and are very fubject to gum.

Of

Of the Management and Pruning of Peaches and Nectarines.

In the management of peaches and nectarines there will be neceſſarily repetitions of ſome things that have been ſaid on the management of apricots and other trees ; but it will (I imagine) be more agreeable as well as uſeful, to have ample directions on theſe ſubjects we are treating of, therefore I ſhall plead its utility as a ſufficient apology for ſuch repetitions.

All peach-trees, proper for planting, ſhould be young and vigorous, have only one ſtem, and never headed down until the ſpring ; as ſoon as they begin to puſh, they muſt be headed down, if ſtrong trees, to ſix eyes, if weak, to four eyes: when headed, open the ground a little on that ſide next the wall, and preſs the tree gently to it, until the top where it was cut oſſ touches the wall.

It

It will be of great advantage to have an equal quantity of frefh cow-dung, and ftiff mould mixed, as thick as common pafte, and put on a thin layer all over the cut part directly; this will prevent the froft, fun, or wet, from penetrating the wound, and keep it from gumming, to which they are fubject. As foon as the young fhoots will reach the wall, they fhould be nailed horizonally; all the fore-right buds muft be nibbed off, and if the feafon is very dry, they fhould have a little water; if the trees were headed down to fix eyes, they will produce at leaft two good fhoots of a fide: the lower-moft of each muft be cut to fix or eight eyes, and the upper ones to four or five the next pruning time. The trees that were headed to four eyes may have good branches on each fide, which may be cut to fix eyes.

In the fpring all the fore-right fhoots muft be rubbed off as they appear; from the fhoots that were cut eight eyes, three or

four

four good shoots may be laid in from each;
and those cut to six eyes, two or three
proper branches may be laid in for wood;
all others must be rubbed off, that those
laid in may have room to grow and the
fruit to ripen: they should be nailed as
they advance in length, for it is very prejudicial to the young wood of peach-trees,
to be blown and twisted by the wind, efpecially where they are strong, as those
that are managed thus will certainly be.

In the autumn they must be pruned,
and shortened according to their strength;
if they have thriven as they should, they
will be all vigorous and in great heart;
but none of the branches should be cut
shorter than six eyes, and the strongest to
nine and ten; they should be nailed as
soon as possible after they are cut. The
next spring they will bear plenty of blos-
soms, and as they are now come to a
pretty good size, they must be carefully
looked over, and all the branches that are.
not to be laid in for wood, rubbed off
while

while young, that there may be no bufinefs for the *knife* at the pruning feafon,
but to fhorten the branches. This is the
beft method to keep peach and nectarine
trees in good order.

PEACHES and nectarines are in general
long in lofing their leaves; autumn is the
proper time to prune them; the trees
now being large, it is neceffary to have a
full view of them before they are cut,
which cannot be done when full of leaves.
It is very prejudicial to the buds to pull off
the leaves when green; for fometimes the
bark is torn, which often caufes the young
fhoots to gum, and fpoils the wood in
many parts of the tree.

A PERSON with a fharp knife will cut
them off very foon, fo that the whole tree
may be feen: they fhould be cut an inch
from the bud, which will foon decay, and
drop off without injuring any part of the
tree. It is a great advantage to prune
peaches and nectarines early, the young

VOL. II.   F   wood

wood then being of a foft nature, it has a large pithy heart, and is liable to be greatly injured by rain and froft, if it happens foon after cutting. When peach-trees are cut early, the days being long, and the fun of great force, the wounds are foon healed, and they are as fafe from froft as if they had not been cut.

ALL trees pufh faft at the extremities, and none more fo than peaches and nectarines ; it is the nature of them to grow in winter, notwithftanding the feverity of the cold frofty weather, in defiance of which they make an early pufh. If they are not pruned in the autumn, they cannot be done with fafety before the beginning of March, and then the frofts are often feverer than they are any time in October. The extremities of the branches which are cut off in pruning, in hard winters, by the end of February are fwelled round ; this is wafting the fubftance of the tree to no purpofe ; neither do the buds

blow

6

blow fo ftrong as thofe that are cut in autumn.

THERE are many objections made by fome gardeners to the pruning of fruit-trees in autumn, but they are in my opinion in general frivolous, and not worth confuting. Let the directions given be carefully attended to and they will all vanifh. As foon as convenient, after the trees are pruned, let them be nailed; all their extremities bending a little to the right and left, from the middle of the tree, never allowing two fhreads to bear the fame way, nor permit the branch to reft againft a nail, for it cankers it.

THERE is many a good tree iniured by being pinched in the fhread; the beft method is to un-nail the whole tree, and then difpofe of the large wood regularly all over the wall, the young wood will fall in properly of courfe.

IN the fpring, when the trees begin to
F 2   pufh,

pufh, they muft be looked over, and all
fore-right and fide-buds rubbed off, leaving none but thofe that are to remain for
next year's wood: this can be done now
with more certainty than when the leaves
are farther advanced; but as many buds
will pufh afterwards, this work muft be
repeated as often as there is occafion, that
is, as long as any fuperfluous wood grows.
If thefe directions are followed, they will
be thriving trees, the fruit good flavoured,
and but little ufe for the knife at the
pruning feafon.

THE thinning of peaches and nectarines
is very material; on the judicious perfor-
mance of this depends the flavour and
fize of the fruit. It is not the largeft fort of
peaches that are the beft, but a large peach
of a good kind is much higher flavoured
than a middling fized one of the fame fort.
If there are a great many fet in clufters, it
is beft to thin them at three different
times; the firft fhould be when they are
as large as a fmall pea; three then may be

3   left

left in each clufter; in twelve days time
another may be taken of, and in a week after that another, always leaving the largeft.
The diftance muft be according to the fize
of the fruit; on the nutmeg kind two
and three inches; on the early forts, three
or four inches; on the largeft forts five,
fix, and feven inches; nectarines do not
require to ftand fo thin as peaches;
three, four, or five inches will be fufficient.

PEACHES and nectarines are generally
thinned with the hand, but that is not a
good method; for where they are fet very
thick, it is impoffible to pull them off
without damaging the ftems of thofe left
on the tree: the beft way is to cut them
off with a fharp knife, leaving a thin piece
of fkin on the tree, which will foon drop:
if they are thus managed, neither the tree
nor fruit is hurt, both of which often happen in the common method of
thinning them. All large vegetables
ought to be banifhed from the peach
borders; then two inches of the prepared

com-

compofition every third year will be
enough to keep the trees in good order.

THE firft time the border is recruited,
the compoft ufed fhould be the fame as in
preparing the border before planting ; but
in a few years it may be neceffary to make
fome change, it being proper to keep the
border to a ftiffifh loam.

The Management and Pruning of Pears.

IN fpring when the buds begin to fhoot,
the tree muft be headed down according to
its ftrength ; that is, four or fix eyes.   As
foon as the branches will reach the wall
they fhould be nailed.   If there are five
fhoots, which fometimes happens (there
will foon be a handfome tree) two of each
fide muft be trained horizontal, that in
the middle upright, that it may be ftrong
againft the pruning feafon.   The tree
headed down to four eyes, if it produces
three good fhoots, will alfo foon make a
good tree : a fhoot on each fide muft be
trained

trained horizontal alfo, and the middle one upright.

If the new headed tree produces only two fhoots, they muft be nailed a little floping to the right and left. If the tree with two fhoots is ftrong, at the pruning feafon cut them to fix or eight eyes, three or four good fhoots may be expected from each of them the following feafon; but if they are weak, cut them to four eyes : there fhould be no more than two fhoots from each allowed to grow. The only way to ftrengthen a weak tree is to lay in little wood, the branches then will make ftrong fhoots. The reafon that fo few proper branches can be laid in from fhoots cut to fix, feven, and eight eyes is, that fome of the buds are fore-right, and fome clofe to the wall; neither of which can be trained with any propriety.

This fummer many fpurs will form on the branches trained horizontal, and on the body of the tree, which fhould not be

F 4   allowed

allowed to grow nearer on trees that bear
fmall fruit than four inches, and on trees
that bear large fruit fix or feven inches.
There will alfo much young wood fprout
all over the old wood of the tree, which
fhould be rubbed off when young, where
fpurs are not wanted ; where they are,
they fhould be allowed to grow until
the wood is hardened, then broke off fix
inches from the branch it grows on, and
in the pruning feafon cut to one eye.

THE next pruning feafon the fame method muft be followed in every refpect as
in the former, until the wall is covered.

THE horizontal branches in fmall pears
fhould be fix inches diftant from one another, and in large fruit eight or nine
When the tree is formed and the fpurs at
proper diftances, there will a great many
young fhoots fprout out of each bunch of
the fpurs, and if allowed to grow large,
will fpoil them : they fhould be all pulled
off when young but one fhoot, which muft
remain ;

remain ; for if they are all taken off, it will caufe feveral of thofe that are forming into buds to fhoot into wood ; this fhoot may be fhortened when the wood is hardened (for after that time no more wood will fhoot) and taken off clofe in the pruning feafon.

If care is not taken the fpurs of pears will grow large and a great way from the wall ; they fhould be thinned to two inches diftance, and every year fome of the longeft cut clean off. It would add greatly to the beauty of the fruit to thin them ; they would alfo be better flavoured ; but this muft be performed after a very different manner from the method ufed in thinning ftone fruit.

Pears drop very much after they are as large as peafe, and there is no knowing thofe that will from thofe that will not, until they begin to fhine, which is a certain fign they will grow : they may be thinned by cutting the ftalk with a fharp knife.

knife.   The borders muſt be recruited the
third year after planting ; for borders
where pear-trees grow are in general made
too free with in growing vegetables, therefore they muſt be repaired in proportion
as the earth is exhauſted. The firſt compoſition that was laid on them ſhould be
the ſame as that made to prepare them for
planting ; but as a few years working and
growing garden ſtuff may greatly alter
their nature, they ſhould be nouriſhed
with ſuch a compoſition as will keep them
as near as poſſible to a good loam, and not
too ſtiff.

The Management and Pruning of Plums.

THE tree ſhould be of one year's growth
from the graft, for if older they do not
produce good wood when headed down,
but luxuriant ſhoots and long rambling
branches.

THE ſpring after planting, they muſt
be headed down to ſix eyes, as moſt kind
of

of plums are very free growers, and if headed too clofe, they fhoot ftrong and are apt to gum. From trees headed down to fix eyes, four or five proper branches may be expected ; if five, the odd one muft be nailed upright, and two on each fide trained horizontal : if there are only four, one on each fide muft be trained horizontal alfo, and the other two inclining to the right and left, keeping the middle of the tree open ; all fore-right fhoots muft be rubbed off as they grow, during the fummer. Next autumn the fhoot that was nailed upright muft be fhortened to fix eyes ; if very ftrong, which it often is, it may be cut to ten eyes ; if the fhoots were carefully rubbed off in fpring and fummer, there would be little ufe for the knife,

The tree that had a branch on each fide fhortened to fix eyes, fhould have four proper branches of a fide, three on each muft be trained horizontal, the other two fhortened to fix or eight eyes and nailed as before directed : the next fummer it will bloffom and bear fruit : there fhould be as many proper branches laid in on each fide as can be done at a regular diftance ; all fore-right and irregular fhoots rubbed off as they appear, and none allowed to grow but thofe defigned for wood. Next autumn the middle branches muft be fhortened to produce more horizontal ones : this fhould be repeated every autumn and in the fpring, as many branches laid in as can be properly got, until the fpace allotted for the tree is full ; but none of the branches trained horizontal fhould be ftopped until they have run as far as intended.

THE diftance of the horizontal branches in fmall plums fhould be from three to four inches, and the fpurs on the branches from two to three inches. On the larger forts the diftance of the horizontal branches fhould be from four to fix inches, and the fpurs from three to fix. The fpurs muft not be allowed to grow

into

into large clusters; they must be thinned
a little every year, and those the furthest
from the wall cut clean off.

SOMETIMES great bunches grow where
the spurs sprout, but that is seldom when
the tree is well managed; when it happens they should be cut clean off with a
chissel, and some of the mixture of dung
and earth spread over the wound to prevent its gumming; there will much
young wood sprout from the place next
spring, one shoot of which should be saved
to procure a new spur. If plum-trees
are managed as here directed, they will
last long in good order and bear a great
quantity of good fruit.

The Management and Pruning of Cherries.

CHERRY-TREES for walls should be
one year old from budding when planted;
they produce much better, and make a
handsomer tree than those that are older.

THEY should be planted in the autumn:

tumn : as foon as the buds begin to pufh in
the fpring they muft be headed down to
four eyes ; when the young fhoots reach
the wall there fhould as many be laid in
as can be properly trained ; that is, thofe
that come from the fides of the tree ; for
all fore-right buds and alfo thofe between
the wall and tree muft be rubbed off.

No tree difagrees fo much with the
knife as the cherry ; if it is rightly managed in the fummer, there will be little
ufe for that inftrument at the pruning feafon : if there are four proper fhoots after
heading, which is often the cafe, as the
buds of cherries are more oppofite to one
another than in moft trees ; one of each
fide muft be trained horizontal, the other
two more upright, but ftill inclining to
the right and left.

If it fhould happen (as it fometimes
does) that two branches can be laid in on
the one fide, and only one on the other,
cut one out, that there may be an equal
num-

number on both fides; for if the tree is
ftarted with more branches on one fide, it
will be impoffible ever to make a handfome
tree: if the odd branch can be nailed upright in the middle to fhorten it for more
wood, it will anfwer well; if the number of branches are even, the two middle
ones muft always be trained a little
more upright, that they may be ftrong a-
gainft pruning time to fhorten and produce
more wood. The trees that have upright
fhoots in the middle muft alfo be fhortened; whatever number of branches are
produced from thofe that are fhortened in
the middle of the tree, an equal number
muft be laid in horizontal on each fide;
if there is an odd branch that can be nailed
upright it will do, if not, let it be cut out;
if the number of branches are even, the
two next the middle muft be trained more
upright to fhorten at the pruning feafon
for more wood. This muft be done every
year until there are as many horizontal
branches as will fill the wall.

CHERRY-

CHERRY-TREES in general produce
plenty of fpurs, but there are fometimes
fpaces where there is only young wood
which grows fingle on the horizontal
branches; this muft be encouraged to grow
until the wood is hardened; they then
fhould be fhortened to fix inches, and in
the pruning feafon cut to two eyes to produce fpurs.

THERE are often on cherry-trees fmall
fhoots that are full of bloffom buds ; thefe
muft be fhortened to an inch and cut to a
leaf bud.

THERE is no tree fo fubject to vermin
as the cherry ; the fpurs generally grow
in thick clufters ; grubs often lodge there
and deftroy the bloffom before it comes
out, and frequently after it blows : to
prevent this, the fpurs fhould not be clofer
on the horizontal branches than three
inches; as they grow in clufters they
fhould never be nearer than half an inch to
one another; that admits a free air all
round

round the bloffom, and there is no harbour
for vermin. The fpurs fhould be kept as
clofe as poffible to the wall, and when they
are thinned, thofe the fartheft from the
wall fhould be cut off.

THIS management will anfwer for all
kinds of cherries but the morella, which
requires a very different treatment : the
morella fhould be planted and headed down
in the fame manner as the others ; what
branches it produces fhould be trained regularly on each fide at as great a diftance
as they will admit of, that others may be
trained in between them the following
fummer. There are no branches to be
fhortened, but as many laid in, as properly can be to the wall, at three and four
inches diftance ; which muft be trained as
horizontal as poffible.

IN the fpring, when they pufh, the bud
next the bole of the tree fhould be laid in
for wood ; and at fix inches diftance, and
continue the fame over all the tree : they

muſt be nailed as they advance in length
The next autumn the whole tree ſhould be
un-nailed, and all the branches ſpread
equally : if there is not room to lay young
wood in the heart of the tree, there may
ſome of the oldeſt, that have bare wood, be
cut out, which is all the cutting they
ſhould have.   They never ſhould be
ſpurred, as they bear the fruit on the
laſt year's wood, which being very ſmall,
ſhould be laid in at three inches diſtance.
They ſhould be carefully looked over in
ſpring and ſummer, and all fore-right buds,
as alſo thoſe that are not deſigned for
wood, rubbed off; for as the branches are
laid in ſo cloſe, if they are ſuffered to
grow rude it ſpoils the fruit and wood for
next year. If they are managed thus, they
will continue long in good order, and bear
great quantities of fruit.

THE morella is always planted on a
north aſpect, being thought fit for nothing
but baking ; but when planted on a ſouth
wall, and they hang on the tree till they
are

are black, it is a fine flavoured fruit and has an agreeable sharpness.

THE management and pruning of pears, plums, and cherries, heading them down, training the branches the first year, and shortening them at pruning time, are so much the same, that it may seem superfluous to say any thing more than what is directed for pears. This might answer in general, but there are some particulars belonging to each sort which should not be omitted.

THE proper distance of all kinds of cherries is from eighteen feet to twenty ; they bear much better so than when they are more crouded.

The Raising, Management, and Pruning of Figs.

THE best method is to raise them from layers, which should be laid in the spring : they will be fit to plant the spring following,

G 2

lowing If a garden pot full of light
rich mould was funk in a convenient place
in the border where a young branch could
be laid, it would foon take root, and
might be cut off from the mother plant
and removed into fhelter the beginning
of winter. The laying the young branch
will no ways impede its growth. As foon
as it has fhot an inch pinch out the heart,
it will produce feveral fhoots, and will be
a fine tree to plant next feafon. The beft
feafon to plant figs is the middle of
March, the roots then are foft and fpungy,
and often mifcarry if planted in autumn,
efpecially if the winter is fevere, although
they fhould be covered.

IF the layer was in a pot, the roots
fhould be pared off, and the ball put into
the ground. The layers that are upon the
mother plant fhould be planted as foon as
poffible after taken off, if in the fame
garden, with all their roots, which muft
be well fpread, for they will be numerous.
If they are to be fome time out of the
ground

ground they fhould be packed in mofs,
and before they are planted all the fmall
roots cut clofe off, and the large ones
fhortened.

ALL the branches of figs fhould be
nailed horizontal, or they will grow too
vigorous and bear little fruit. The proper
time to prune them is in October. If the
froft has not deftroyed the leaves, they
muft be cut off; the little they bleed at
this feafon will not be of prejudice to
them.

THE only pruning they require is to
cut out fome of the ftrong old wood, to
make room for young fhoots to be regularly trained over all the tree, for on
them only fruit is produced. As foon as
pruned the late figs muft be pulled off,
and all the branches nailed clofe to the
wall, and covered before the froft grows
fevere.

IN fummer, all the fore-right branches

and small wood that is not wanted should
be pulled off when young, for if they are
permitted to grow strong they bleed
much, which is a great detriment to the
tree and fruit.

The Management and Pruning of Apples.

IT is uncommon to plant apples against
walls, unless on north aspects, where
sometimes a nonpareil is placed to fill up
a vacant spot.

NONPAREILS and golden pippins
planted on the south and south-west aspects
are superior in flavour to many peaches,
and preferable to most plums; no good
garden should be without a few trees in
a proper aspect.

APPLE-TREES planted against walls
should be young and vigorous, and have
only one stem free from canker, to which
distemper they are subject.

THEY

THEY ſhould be grafted on free ſtocks, that is, ſtocks raiſed from the ſeeds of large ſour apples, or crabs; but I think crabs make the fineſt trees.

THE proper time for planting is in autumn, unleſs the ground is ſubject to wet in winter; in that caſe the ſpring is preferable. They muſt be headed down in ſpring, and treated in every reſpect exactly the ſame as pear-trees: although their fruit and foliage is very different, they are much of the ſame nature; the only thing in which they differ from pears is, that they are apt to have their ſpurs grow too cloſe, which cauſes green worms to lodge amongſt them and eat the bottoms of the bloſſoms; this is often erroneouſly called a blight.

DWARF apples, round the quarters in a kitchen garden, ſhould be planted at ſix, ſeven, or eight feet diſtance, if on French paradiſe ſtocks; but if on Dutch paradiſe ſtocks they may be planted at eight,

nine,

nine, and ten feet diſtance ; but theſe
ſhould never be planted facing ſouth walls,
even if the borders are twelve feet wide,
and the walks of equal breadth. If
apples on Dutch paradiſe ſtocks are
planted on the borders of the quarters
behind the north aſpect, and allowed to
grow twelve or fourteen feet high, and
the garden is of an eaſy deſcent, it will form
an agreeable view in ſummer from the
walks on the ſouth ſide of the garden.

WHEN Dutch paradiſe apples are planted
to grow twelve or fourteen feet high,
they muſt ſtand at ten feet diſtance, and
be trained to fill up the whole ſpace ; but
in ſo looſe and eaſy a manner that they
may appear natural, and have no reſem-
blance of a hedge. When they are allowed to grow ſo high, there is ſome
difficulty in keeping them from being
naked at bottom ; they muſt not be trained
thick at firſt, and have ſome branches
ſhortened every other year, to keep young
wood in the middle and bottom of the
tree ·

tree ; the fpurs muſt not ſtand too cloſe, nor allowed to grow in too great cluſters.

IF the apples were planted in autumn, and were fine young plants with three or four good ſhoots, they muſt be headed down in ſpring, the two lowermoſt to ſix or ſeven inches, the others to about ten or more, according to their ſtrength.

THEY muſt be looked over in ſummer, and no tufts of young ſhoots permitted to grow but ſingle ſhoots, and they at a proper diſtance, a foot and eighteen inches.

IF they have thriven well, there will be four or five good ſhoots at a proper diſtance on each of thoſe that were headed down ; and muſt be ſhortened at the pruning ſeaſon to produce more wood, that the tree may be properly fur-niſhed as it advances in ſize.

THERE is no certain rule for ſhortening the

the branches (as in wall-trees) two or
three of thofe next the root muft be cut
the fhorteft, to fill the bottom of the tree.
The top will always be full enough, and
fhould be thinned in the pruning feafon,
that the bottom part of the tree may have
free air to perfect its wood. and ripen its
fruit. Thofe on Dutch paradife ftocks
fhould be treated in the fame manner,
only they fhould be kept thinner, as they
grow much taller.

ESPALIERS are in general laid afide;
they require a great deal of labour, and
are fo ftiff and formal that there are few
good kitchen gardens now have any.

THEY hide the quarters in fummer,
although they render the kitchen garden
more agreeable to walk in; but this is
not to our prefent purpofe. The kitchen
garden not being a fit place for walking,
unlefs to thofe who are fond of its product,
and feeing the things growing and brought
to perfection; if the garden is kept clean

6   and

and neat, the growing of vegetables is. what it is defigned for, and ought not to be hidden.

If there are neat frames made, it is a great expence at firft, and they are formal and ftiff, in a few years they begin to decay, and then muft be patched and mended; in winter they are very unfightly.

Those made with pales are the beft; for although they are to repair every year, after the fecond or third, it is a trifling expence, as all the work is done by the common labourers, even if the pales are bought at the deareft market, which feldom happens, as they are to be had in moft country places. In fummer, when the trees are in leaf, thofe efpaliers made with pales are the handfomeft, becaufe not fo clumfy; and in winter the frames have no great beauty.

The trees for efpaliers, whether apples or pears, fhould be young plants, having
the

the same qualities as those for walls, and be treated in every respect the same, in planting, heading down, pruning, and training.

ALTHOUGH there are dwarfs, or espaliers, all round the quarters of a kitchen garden, it will be far from being sufficient to serve a family. Thus few small families have large kitchen gardens, the apples for winter, summer pears, and other common fruits that can grow even in a large kitchen garden, will be inferior to the consumption of a small family. A small family should have in some convenient place a few baking apples, summer pears, some of the common baking plums, a tree of baking pears, some damsons, and filberts on standards.

THE above are necessary in all families; but when they are planted in the kitchen garden they entirely spoil the herbage. Where the family is large there ought to be a large orchard, in which there should be

a quana quantity of all the above fruits; to which should be added, walnuts, chesnuts, mulberries, and almonds. In the north they seldom succeed, but in the southern countries they bring great crops.

CHAP.

C H A P. XIII.

Of the *Ananas*, or *Pine-Apple*.

THE culture and management of the Ananas, or Pine Apple, is brought to fo great perfection cf late years, that it may be imagined it cannot be now improved, and may feem quite unneceffary to fay any thing refpecting its propagation; but as there many things concerning them materially, which are unknown to the generality of thofe concerned in pine ftoves, for their inftruction I fhall give eafy and fafe directions for raifing the plants, and bringing them to fruit much fooner than by the common method.

THE common method with the crowns is to cut off the foft pulpy end that is

twifted

twifted out of the fruit, and to pull off a few of the under leaves, and lay them fome time in a fhady dry place to harden ; then plant them in fmall halfpenny pots, and plunge them into a bed of tanner's bark, made ready in the breeding-ftove for that purpofe.

It is certain that a plant lying fo long in drying, after it is deprived of its nourifhment (the fruit) muft be a detriment to it, and greatly retard its growth, for the leaves become foft and languid like any other plant that is much dried in a funfhiny day; and were it not for the nature of the leaves being ftrong, and of a very different texture from woody plants, it would fhew its wrong treatment as much as thofe would were they treated in the fame manner.

The cutting off the foft end that is twifted out of the fruit deprives the bottom of the plant from ever pufhing out a root, for it remains to the laft juft as it

was

was cut off, and the pulling of the leaves from the bottom deſtroys moſt or all off the ſmall little knobs that are formed between the leaves, and are the rudiments of the roots, all of which grow when treated in the method here directed.

Those alſo that are managed in the old method are very ſubject to rot if taken off late in the ſeaſon from the fruit, as then they cannot be ſo well dried; whereas thoſe that are planted in the new method are in no danger if taken off in the middle of winter, and they are no trouble, nor do they require any place to be made on pur-poſe for them.

The fruiting pines are, or ought to be, put into freſh heat in the ſpring, which will be a fine moderate warmth by the time there are any pines fit to cut ; and as there is always near a foot in front of the bed between the flue-wall and the pots that ſtand in the foreſide of the pit, it will be ſpace ſufficient to hold many

6

crowns

crowns, though of no ufe in the prefent
method, neither fhould there be any other
ufe made of it ; for if any pots were to be
plunged there, unlefs they were very fmall,
they would deprive the forefide row of
heat; and if the plants in them are of half
the height of the pines, they prevent the
free circulation of the air amongft the
pots, which would be a great detriment
to the fruit.

As foon as you can procure the crowns
(the fooner the better after being feparated
from the fruit) make a whole in the bark
with a fetting-ftick two inches deep, then
put in the crowns, and make them faft in
the bark.

WHEN the crowns have been fifteen
days in the bark they fhould have a little
water once a week, and may be planted
within two inches of each other, and the
crown that is next the fruiting plant may
be placed within two inches of it, fo that

VOL. II.   H   there

there may be three rows of crowns all along the foreside.

It will be the best method to carry them all on as the crowns can be got, for after they begin to grow it will be of great service to them to be sprinkled over with water twice a-week, as it will keep the tan moist; but this cannot be done where there any plants lately planted, for if they get any water before they begin to grow it rots them. If there are any crowns that are taken off in winter, they may be planted into the bark in any place amongst the plants where there is room and air, and not very hot, where they may remain until there are some young plants to shift, when they may be taken up and potted. The crowns that are taken off last in the general crop must not be planted in the front of the fruiting pit, but be disposed of in the same manner as those taken off in winter; for the house must be all removed as soon as the fruit is cut, so that they would not have time to get root.

BEFORE any of the pots are removed raife the crowns out of the bark carefully with a ftick, fo as to break none of their roots. The pulpy part that was twifted out of the fruit will be all rotted off, and the bottom will be fmooth and found, and in good condition to pufh roots.

THERE will be many of the under leaves alfo rotten, which muft be pulled off; then there will be a good ftem, hard and found, with many knobs for pufhing roots; befides there will be many fine roots which have ftruck while they were in the bark, which will receive no check in being removed. Cut the end of four or five rows of the fmall leaves round the bottom, that they may decay before the next fhifting in order to get more ftem to pufh out roots, for the longer the ftem is the more vigorous will the roots be, and the plants will grow very faft. Plant them into large halfpenny pots, taking care to lay the roots fmooth, and plunge them into a moderate heat, giving them a little water

H 2   the

the next day, and they will require no more for some time.

THEY should be shifted again into penny pots before the roots are much matted round the pots, which will be in about six or eight weeks. None of the roots must be disturbed, but taken carefully out of the small pots and planted into the larger ones, having first put two inches of mould into the bottom of those pots. Let the mould be carefully put round the ball, so as not to break it, yet to make the plant fast, so that there be no vacancy. Plunge them into a tolerably good heat, and give them a little water; but they should have no more for a week, for pines should never have much water after shifting, until they begin to grow again.

THIS work may be done in any of the winter months as safely as in the summer. A moderate heat, a little water, and fresh air should never be omitted, if the weather is ever so cold, for it is against nature to

imagine

imagine a plant can live and be in good health without frefh air. Although pines do not fo foon difcover the want of it as many other plants that are more fucculent or foft-leaved, yet they equally fuffer for want of it.

IF the plants have had no misfortune and are thriving, they will require to be fhifted into three-halfpenny pots in the month of May, when they will be ftrong plants, and be in no danger of what the gardeners call *running* (fruiting).

I CANNOT fay pofitively that this fhifting will entirely prevent it, as perhaps a few may fruit; but I can fay with truth, that there will not be twenty in five hundred, which is next to nothing. Notwithſtanding this, thofe that have had a check by cold, or being too much heated, will be liable to fruit.

IT fometimes happens that old plants will not fhew fruit at the proper time,

though

though they appear large and in good health. This often happens when the roots of the fruiting plants are burnt all round the fide of the pot and go no further. There are roots fufficient to fupport the plant for a time ; and if they do not get too much water at that juncture, having a good heat, they will foon pufh frefh roots, and keep growing inftead of fhewing fruit, which they feldom do until the pots are full of roots. This is the reafon why fome plants are fo late in fruiting; and is alfo a convincing proof that fhifting young plants, before the roots are matted round the pots, is a good method to prevent their fruiting.

The fuckers fhould remain on the ftools, or old plants, after the fruit is cut, until the whole crop is finifhed ; then they fhould have a good deal of water, efpecially if there is a moderate heat in the bark As the old plant has nothing to feed but the fuckers, they will grow to a large fize before all the crop is cut, which will
be

be about the end of September, if the plants fruited at a proper feafon.

ALTHOUGH the tan is in feeming good order between the pots, yet underneath it will be dry, hufky, and mouldy, if the plant has had no water for fome time before the fruit be cut; therefore as foon as the fruit is off it fhould have a large quantity of water, which will run through freely, moiften the bark, and caufe it to ferment afrefh, which will greatly encourage the growth of the fuckers. If they are low in the pot they will ftrike roots and be fit to pot as foon as they are taken off, which may be done with fafety, if there is a proper heat to plunge them in; for as they have been long on the old plant their ftems will be hard and dry.

As foon as they get roots they are almoft ready to drop off, fo there will be a very fmall wound where they are taken off the old plant, and no danger of rotting; but if there is only a few that have

H 4   roots,

roots, it will not be worth the trouble of potting a small number. Cut the roots off and plant them with the unrooted suckers, as hereafter shall be mentioned. If the roots are not cut off they are apt to rot when planted in the bark, and are detrimental to the young roots when they push.

SOME time before the suckers are intended to be taken off, there should be a place prepared in the breeding-stove for their reception : the bark should be trenched over, and as much new bark added as will make a moderate heat, and at the top there should be at least a foot of old bark without any mixture of new ; and when very dry, it should be watered until it is moist.

THIS should be done seven or eight days before the suckers are taken off, that the heat may rise 'before they are planted ; for if the bark be quite cold they will be in danger of rotting ; and, which is full as

bad

bad as if the heat was to be violent and
fcorch the ftems, which would retard the
plants three months, if it does not quite
fpoil them ; but if only a fmall quantity
of new bark be added, there will be little
danger of its being too hot.

As foon as the bark is a little more than
milk-warm, take off the fuckers from the
old plants and plant them immediately
with a fetting-ftick in the prepared bed,
juft as they are taken from the ftools. They
may be planted in rows at fix or eight
inches diftance as they are in largenefs,
and four inches afunder in the row. They
fhould be four inches deep in the bark,
and if made faft it will be of great fervice
to them to be planted that depth, for it
will rot the fmall leaves round the bottom
of the plant, fo that they may all be taken
off when they are potted. Their being
planted in this manner a fmall fpace will
hold many plants, which will give an opportunity of preparing the beds for their
reception when they are potted.

THEY

THEY muſt have no water for twelve or
fifteen days after they are planted in the
tan, by which time they will have begun
to puſh roots ; then they ſhould have a
little water, if the weather is warm, twice
a week ; but if dull and cold, once a week
will be ſufficient. When they begin to
grow freely they muſt be watered frequently ; and it will greatly encourage
them to ſprinkle them all over once or
twice a week if the weather is fine, and a
good deal of ſun ; but in dull days they
require none, as the air is then ſufficiently
moiſt.

IN ſix weeks they will be fit to pot,
therefore muſt be carefully taken out of
the bark with all their roots uninjured ;
then take off the dead leaves ; but care
muſt be taken to pull them no higher on
the ſtem than it is of a dark brown, for
if they are taken off until the ſtem is
white, they are in danger of rotting. All
the ſmall leaves round the bottom that are
not decayed ſhould be ended, to make
them

them rot before the next fhifting, that they may be pulled off to get a good ftrong long ftem to pufh vigorous roots.

As many of the fuckers will be very large, and have made roots five or fix inches long, they muft have penny pots at their firft potting ; nor fhould they remain longer in the bark than to make roots of that length, for if the roots are longer they muft be either cut off or twifted round the pot, or put into pots too large for them, all of which are very detrimental to their growth. If there fhould be a root too long to be laid even, it is better to cut it off at a joint than to twift it round the pot, or to plant the fucker in too large a pot.

By allowing the fuckers to remain fo long on the mother plant, they grow much fafter than if they were taken off as foon as the fruit was cut: they alfo harden in the ftem, and as they grow hard, they alfo form fmall knobs on the fides of
their

their ſtems, which puſh ſtrong roots immediately on their being planted in the
bark, and the ſtem, by being hard, is in
no danger of rotting.

By being planted directly when they are
taken off, they loſe no time, and are not
retarded in their growth, which they do
when taken off and laid to dry, which is
the common method of treating them.

Before they are planted the leaves are
ſhrivelled, and it is ſome weeks after being potted before they recover their proper
poſition. The ſmall knobs that are on the
ſtems, which are the rudiments of the
roots, are all dried and loſt ; freſh knobs
muſt therefore puſh before there can be
any growth in the plants, which makes
it evident that the method here directed
forwards the plants almoſt a year, and is
much eaſier and more certain than the
common method that is in general practiſed.

Before the ſuckers are taken out of

the bark, there fhould be a proper bed
ready, and the heat pretty ftrong; for as
they are to be planted in penny pots, and
the roots not fuppofed to reach the fides of
the pots (for reafons before given) they
will grow much better if the heat is pretty
good.

As foon as they have been plunged a
day give them a little water to fettle the
mould to the roots, and they will not require any more for ten or fifteen days;
after that time they fhould be watered at
leaft twice a week, and if the weather is
dry and hot, it will greatly encourage them
to fprinkle them with water once a week,
unlefs it be late in the feafon, when the
nights are long.

THEIR being fhut up without air fup-
plies the leaves with moifture without
fprinkling, as the houfe will be fufficiently damp all that time; neither fhould
there be any water allowed to fall into their
hearts in the winter; but it is an advantage

tage to pour it in amongſt the bottom
leaves, in watering, even in winter, as it
waſhes out all the dirt and naſtineſs which
is very apt to lodge in their bottoms, and
ſometimes rots them off cloſe to the pot,
or decays the plant ſo that the fruit ſtem
becomes crooked, and the fruit grows de-
fective on that ſide.

As ſoon as the roots are come round the
pot, ſo that they can be taken out without
danger of breaking the ball, they ſhould
be ſhifted into large pots. Eight or ten
weeks is the uſual time allowed them for
ſhifting, but you may judge to a certainty by turning the plant out of the pot
carefullv, at leaſt ſo far as to ſee in what
condition the roots are ; for they ſhould
not be allowed to remain till the roots are
matted, for then there is ſo much ſmall
ſoft woolly ſtuff grows amongſt the roots,
which is abſolutely neceſſary to be taken
off, and in doing it many of the beſt and
ſtrongeſt roots may be broke, and the ball
often greatly diminiſhed, which is wrong,
and retards the growth of the plant.

IF the woolly ſtuff is not taken off it moulds, and very often deſtroys the roots after there is freſh mould put round them, and it prevents the roots from coming to the ſides of the pots.

THEY ſhould be carefully taken out of the penny pots and planted in three-half-penny ones ; and if the heat is declined, a little freſh bark ſhould be added and work-ed up with the old, which will renew the heat. This work may be performed ſafely in any of the winter months.

THIS ſecond ſhifting ſhould be about the beginning of December or February, according to the time the ſuckers were taken off in autumn.

THIS ſhifting keeps them growing all winter, and greatly prevents their fruiting ; for the reaſon of many young plants fruit-ing is, a want of earth to ſupport their roots, want of moderate heat to keep them growing, and of water to keep them moiſt;

for

for if they are in small pots, the pots are soon full of roots, and must be often watered.

IF there is not a good heat they are in danger of being rotted, and if not watered they are sure to fruit; but if they are shifted and kept growing, and have a good deal of air, they will be fine plants, and produce much better fruit than those that are drawn tall, and to the look seem much finer.

ABOUT the middle of May the roots will again be all round the sides of the pot; they then should be shifted into two-penny pots in the same manner as before directed; but it will be necessary to examine the roots, to be certain of their being in order for shifting.

As the weather will now be warm, if a tolerable season and a good deal of sun, as soon as they have been plunged a day they should have a little water, and in about

six

fix days they will require a little more, and after that they muſt be watered frequently, according to the warmth of the weather.

WHEN the bark becomes dry at the top, it loſes its heat, therefore it would be better to water them all over with a watering-pot and roſe often : this keeps the bark in a moderate ferment, and the plants grow freely. They ſhould have a good deal of air ; for if drawn at this ſeaſon they never will bring good fruit.

WHEN two or three dull days ſhall happen together, they ſhould have ſome air every day, if it were only for an hour, although the houſe is not ſo hot as it ought to be ; for as at this time of the year there is no fire, in dull weather the ſtove will be very cold, and it may be imagined there is no occaſion for air ; but it is very neceſſary ; and even in froſty weather, when there are fires, if the ſtove is cold, yet it ſhould have freſh air, which

VOL. II. I ſhould

fhould be admitted by the back-doors into the fhed.

AFTER this fhifting they may remain longer than they did after any of the former, as they have now a good large ball of earth to fupport them, and may now fafely grow in the fame pots to the middle of Auguft, when they may be planted in the pots they are to fruit in ; at which time they fhould have all the fmall ftraggling roots cut off clofe to the ball.

THERE is a great advantage in putting the fruiting-plants early into their fruiting-pots, as there will be much fine weather after, that time (the middle of Auguft) they may have a good deal of air while they are getting new roots ; and as the bark will be warmer at this fhifting than after any of the former, there will be an opportunity, when the weather is fine, of giving a good deal of air, the plants will thrive, grow ftiff and ftrong, and produce good fruit ; and as the pots
will

will be full of roots early in autumn, they will feldom fail of fruiting at a proper fea-fon.

BEFORE the plants are fhifted into frefh pots the bark-bed fhould be worked to the bottom, all the rotten tan taken out, one third of new added, and it fhould be well mixed with the old: this thorough ftir-ring is neceffary at this feafon, that the bark may keep a moderate heat all the winter; but it frequently happens that the heat is violent for fome time, and if the pots were plunged up to the rims it would deftroy all the roots (and fometimes moft of the ftems) which would be a great de-triment to the plants, and would be a long time before they recovered ; for as long as the violent heat continues they will pufh no frefh roots ; and if any quan-tity of water is given during that time, they are in danger of rotting.

WHEN a misfortune of that kind hap-pens, the beft method is to take the

plants

plants out of the pots, cut off all the rotten roots, and what is ſpoiled of the ſtem ;
plant them into three-halfpenny pots, and
plunge them up to the rim into the ſame
heat, and they will ſoon puſh freſh roots.

As ſoon as the roots are come round the
pot, ſo as to be taken out without breaking the ball, they ſhould again be put into
the fruiting-pots : this will retard them,
but not ſo much as if they had been left
in their old pots with their rotten roots,
and their fruit will be better and larger.
They ſhould have little water for ſome
time after they are put into the little pots ;
for as long as the bark is very hot it will
cauſe a great moiſtneſs all round the pots,
which will be ſufficient for them for ſome
time, as they have no roots; but when
the heat begins to decline, and the roots
are advanced to the ſides of the pots, they
ſhould have a good deal.

To prevent accidents of this kind, the
pots ſhould be plunged only half way at
firſt ;

firſt; and when the leaf riſes, which will be in four or five days, if it is very great, they may remain in that poſition for a month or two; but the time muſt be juſt as the violence of the heat continues.

As ſoon as the heat declines the pots ſhould be taken out, and the bark ſtirred eighteen inches deep, and the pots plunged to the top. If the bark was ſtirred any deeper it would cauſe the heat to be as great as it was at firſt, which would be a great loſs, as the pots could not then be plunged to the rims; for they always puſh fine ſtrong roots juſt at the neck of the plant, and when the pot is only half plunged they want heat to encourage them. The heat after this will remain moderate all the winter, and the plants be in a ſlow growing ſtate, ſo as to be in vigour to ſhew their fruit.

THERE is another method to prevent burning, which may be done with leſs trouble; for it is not always that the tan

heats

heats so violently, and there is little danger of its ever doing so, unless the old bark is very dry and not much rotted ; when that is the case, a very little bark will make a great heat, especially if the bed is stirred from the bottom. The plants should not be shifted into new pots, but removed into some convenient place until the bed is made ready, and then set level on the bark for ten days, by which time the heat will be come to its greatest height : he plants may then be shifted, and plunged half way, or to the rims, as the bark is in condition. While they remain on the bark they will require water every day, if the weather is hot, and a good deal of air. They should not be crouded too close, but stand at the same distance as if they were plunged ; for when they stand thick the leaves get a wrong position, and are long before they come right again (if ever) which makes them very unsightly.

WHEN the heat is become moderate, and the plants plunged to the rims, they may

may remain in that condition to the end of February, giving them moderate waterings, and as much air as the weather will permit, keeping the heat of the houfe to a moderate temperature.

THERE are many large and fine plants fpoiled by keeping the houfe too hot at the time of their fhewing fruit, befides it caufes the fruit-bud to be fmall, the ftem to run to a great height, and to be fo very weak that it has not fubftance fufficient to feed the fruit.

THE beginning of March the plants fhould be all taken out of the bed, and a little new bark added to the old, and worked up fo as to be well mixed, to make juft a moderate heat. At this feafon, when the pots are full of roots, if the heat was too ftrong it would fpoil the whole crop; for fhould the roots be now burnt, there is no recovering them.

THE dead leaves fhould be pulled off all
I 4   round

round the bottom, and the earth ſtirred a little all over the top of the pot, and raiſed cloſe round the neck of the plant, and an inch of very rotten dung ſpread all over the top of the pot ; for as the earth in the pot has fed the plant ſo long, it muſt be much ſpent ; but the dung being waſhed down amongſt the roots with the water, it will add freſh vigour to the plants, and cauſe the fruit to grow freely.

THEY ſhould now be plunged to the rims, and have a good quantity of water, not too much at a time, but often ; they ſhould alſo have a good deal of air, and when the weather is hot and dry it will greatly encourage the fruit, if you ſprin-kle them all over with water twice a week, which may be continued until the fruit is come to its full growth ; after this they ſhould not have one drop. The fruit then cuts hard, dry, and fine, and is of a delicious high flavour ; but if watered af-ter that time, they are full of a watery juice, flat and inſipid. As they ſhew at
different

different times there will be many full grown, when there are some not come to half their size, therefore it will be necessary to give these water, but the sprinkling should be given over.

THE crowns which were shifted in May will require to be shifted again the beginning of July; but as it will be impossible to get them strong enough to bring large fruit this season, it will be better to keep them over the year, and if well managed they will be strong, fine plants.

IF they were to be managed in their shiftings as is directed for the suckers, most of them would fruit, but they would be much smaller than the fruit of the suckers; for if the fruiting-plants had no misfortune, the suckers taken from them in the autumn will be larger when taken off than the crowns are now. The stems of of the suckers are also prepared for rooting before they are taken off the mother-plant, and grow much faster than the crowns.

It

It will be much better to keep the crowns growing for another year, as they will then be ftrong, fine plants, and will produce large good fruit.

IF it is intended to keep the crowns growing when they are to be fhifted in July, cut off all the roots round the ball, and leffen it fo much that there will be an inch of frefh mould all round the ball, when planted in the fame pot they were taken out of. Plunge them in moderate heat ; and two days after they are plunged give them a little water to fettle the earth ; but they fhould have no more for ten or twelve days, by which time they will begin to pufh frefh roots, and will require to be refrefhed with water according to the heat of the weather, obferving to give them plenty of air. They will not look fo green and pleafant to the eye, as thofe plants that are kept with little air, whofe leaves are long, fmall, and thin, and drawn quite upright : Their leaves will be fhort and broad, and grow almoft horizontally.   5

THE middle of September they should
be shifted into pots a size larger; but
none of their roots cut off. The ball
should be put in whole as they came out
of the other pots, and plunged into moderate heat. Little water will serve them
in winter, having a good ball of earth to
support their roots, and if they are kept
in a moderate heat they will grow all the
winter.

ONE principal reason why many of the
crowns fruit in the spring is, their being
planted in small pots to stand all winter,
and are generally plunged into a too great
heat in the autumn, which presently fills
the pots with roots: the strength of the
mould is soon spent, and then they have
nothing but water to support their roots,
which if given in any great quantity, and
the heat declining, they grow yellow, and
sometimes are rotted. If they do not get
water, their being dry stops their growing, and makes them set for fruit, and as
soon as they are put into larger pots and
fresh heat in the spring, they shew fruit
directly.

THE beſt method to prevent their fruit-ing is to give them only a moderate heat, and to keep them juſt moiſt; and if the heat declines any time in the winter, the plants ſhould be taken out of the bed and a little freſh bark added: this will renew the heat milk-warm, the degree of which is ſufficient for them at this time of the year.

THE beginning of March they ſhould be taken out of the pots, many of the roots cut off, ſo that there may be a ſufficiency of freſh earth round the ball when they are put into penny pots; being abſolutely neceſſary to ſhift them at this time, and to give them freſh earth to promote their growth, as that in which they have been in all the winter will be much ſpent.

IT is alſo neceſſary to plant them into leſs pots; for if they were to be put, at this ſhifting, into pots a ſize larger than thoſe they are now in (as the ball ſhould be no more broke after this) the pots they

muſt

muſt be planted in to fruit, would be too
large, and take up too much room in the
ſtove to no purpoſe; beſides, the plants
would not thrive ſo well were they in ſuch
large pots. A juſt proportion between the
plant and pot is neceſſary to be conſidered
for the encouragement of its growth.

About the middle of May they will require ſhifting into two-penny pots, as before directed; the roots ſhould be examined, and if neceſſary, ſhifted ſooner; for
the roots muſt not be allowed to mat, neither ſhould the ball be broke.

As they now will have a good ſupply of
earth to nouriſh their roots, they ſhould
remain in theſe pots until the middle of
Auguſt, when the ſuckers of a year old
are ſhifted for fruiting; and then put into
the three-penny pots, and plunged in the
ſame bed. As they will be much thicker
of leaves than the ſuckers, it will be proper to mix them; they will look better,
and

and there will be more air amongst the pots
than if the crowns were to be plunged all
together.

IF they are properly shifted and managed as here directed (and meet with no
accident) they will be strong, short, stiff
plants, with broad leaves growing almost
horizontally, and will shew fine fruit-buds
with short stems, and produce a large
rich-flavoured fruit.

MANY admire crowns, as they make the
handsomest plants ; they are much thicker
of leaves than the suckers, and more
sightly ; but if the suckers are properly
managed they will produce as good fruit
at one year old as the crowns will at almost
three, so that there are two years saved,
besides the expence and trouble.

SHOULD the plants which are in fruit
meet with the misfortune to have their
roots scorched, which may happen to the
most

moſt ſkilful and careful perſon, if not very
much damaged, they will puſh freſh roots
and ripen their fruit, although it will not
be ſo large; neither will the flavour be
high, the ſuckers will be very ſmall, and
not fit to produce fruit at a year old. For
fear of an accident of that kind, it will be
proper to have ſome crowns every year
coming on of a year old, to prevent your
being deſtitute of good fruit.

WHEN a misfortune happens to occaſion
the ſuckers to be very ſmall, they will alſo
be very late, and ſhould remain on the old
plants as long as it can be done with convenience, as they will grow much faſter
on the mother-plant than they will after
taken off, beſides the advantage of their
ſtems being hardened. They ſhould be
planted in the bark when taken off, as before directed, and managed in all reſpects
as the crowns that are kept till they are
three years old.

IT

IT may be objected to this management, that the frequent fhifting will retard the growth of the plants, and that they would be much larger if they had been lefs fhift-ed; granted: they would have been much taller, but would not have half the fubftance. A fhort-leafed, ftrong plant will produce much larger and better fruit than thofe that are longer leafed; they indeed are more fhewy to the eye, being of a darker green, and look more vigorous; but they produce fmaller fruit : neither are they fo high-flavoured as plants tnat have more air.

HAVING gone through the manage-ment of the plants in all their different ages, I fhall now proceed to give fome directions about the moulds they fhould be planted in at the different times of of fhifting, which has fucceeded very well with me for many years.

To make good mould for pines, there fhould

should be in readiness the following simple moulds, which ought to be compounded some time before they are wanted. Good light loam surface, that has laid in a heap till the grass is quite rotten, and has been turned several times; rotten dung that has been turned until it is become as small as mould; light rich wood-earth that has been turned four or five times; rotten bark from the stove, sifted, that has laid a year, and been turned several times; and good sharp sand, and if the sand was turned once or twice it would be the better.

THE turning of the different moulds, before they are mixed, is of great use to them, as it meliorates all their different particles, and makes them fit for mixing. It should be done in the winter, when there is hard frost and no snow, for snow is very bad for all compositions that are not made of long straw, leaves, &c. and to them it is of great service. The frozen parts should be turned into the middle,

and the ſmall to the outſide, that it may
get the benefit of the froſt and air to mellow it; then make it into ridges of four
or ſix feet broad at bottom, as there is
room, and bring them to a ſharp point at
top to throw off the rain, which, like the
ſnow, is very detrimental to all compoſi-
tions. In that form it is alſo more convenient for turning, and at every turning,
the bottom ſhould be thrown uppermoſt.

HAVING all theſe moulds, tan, dung, and
ſand prepared, a proper compoſition may
be made of them for pine-apple plants of
all ages.   It ſhould be made up ſome time
before it is uſed, and turned ſeveral times,
ſo that the different kinds may be well
mixed and incorporated.

MOULD for crowns and ſuckers at their
being potted when taken from the tan-bed,
where they were planted to ſtrike roots.
——Three wheel-barrows of light wood-
earth, one barrow full of ſifted tan, one
barrow of loam, half a barrow of ſand
and

and one barrow of rotten dung; this is the proportion: the quantity muft be accord ing to what is wanted; but it would be much better if there was as much mixed at a time as would ferve for two or three years; for the longer all compofitions lay, the richer and better they are; and the oftener they are turned, it adds to their fertility; for it would be poffible to ferti lize very poor foil by throwing it into long ridges and often turning it, which muft be owing to the nitrous particles that are imbibed by its loofe furface.

Mould for the firft fhifting.——Two barrows of wood-mould, two barrows of loam, one barrow of dung, one barrow of tan, and half a barrow of fand.

Mould for the fecond fhifting.—Three barrows of loam, one barrow of wood-earth, one barrow of dung, and half a barrow of fand

Mould for plants that are to be planted
into

into their fruiting-pots; or plants that are large, and may want fhifting out of courfe.——Four barrows of loam, two barrows of wood-earth, two barrows of dung and one barrow of fand.

THE crowns that are to be fhifted in May, and to have their toots cut off round the ball, fhould have mould a little dif ferent from any of the former, that they may grow ftrong, and their leaves broad and thick. I have found the following compofition anfwer well: Two barrows of loam, two barrows of dung, one barrow of wood-earth, and a barrow of fand. This will alfo do very well for them when they are planted into larger pots to ftand the winter.

IN the fpring, when the ball is reduced, they fhould have the following compofi- tion: A barrow of loam, a barrow of wood-earth, a barrow of dung, and half a barrow of fand; and in their other fhift- ings the mould before-mentioned will do for them.

WHEN fruiting-plants are damaged in their roots, fo as to render it neceffary to take them out and plant them into little pots, the following compofition will make them foon pufh very fine roots, and in a little time they may be placed in their fruiting pots again   One barrow of fifted tan two barrows of wood-earth, one barrow of dung, and half a barrow of fand.

I DO not pretend (as I faid before) to affert that thofe moulds which I have mentioned are the very beft compofitions for pine-plants, or that fome other mixtures may not be as good, if not better; but I advance nothing upon credit, nor direct any thing but what I have practifed with fuccefs for many years, and have always found thofe compofitions anfwer to my utmoft wifhes.   The plants will thrive well, and produce large fruit.

THE culture of the plants has been treated of at large from the crowns being
K 3   taken

taken from the fruit, and the suckers from
the mother-plants, to their being put into
order again for fruiting. I flatter myself
that those who follow my directions with
accuracy, will find that nothing has been
advanced but what is practicable, and will
answer.

THERE may be, and daily happen, accidents that prevent the success of the best-
concerted schemes, and the best directions
may be frustrated by a trifling accident or
neglect, so that no scheme or directions
should be given up as impracticable or improper, because they do not always succeed.

ALL those that are the least acquainted
with gardening in general, and hot-houses
in particular, are sensible that under the
care of the most skilful practitioners they
are liable to many misfortunes, although
very sensible how and when they happen,
but too late to remedy them for a season.

I KNOW

I KNOW of none that has been of worfe confequence than too much heat; for, by what I have feen, ten places out of fifteen, where plants have been in bad condition, proceeded from too much heat and too little air, for they do not require fo much as is generally imagined; and although the roots feem hard and dry, they are foon burnt, after which it is a long time before they begin to grow again.

WHEN the roots are burnt, the mould muft fuffer; it is much the beft way to throw it away and frefh pot them. The old burnt roots that are all through and round the mould, muft be detrimental to the young roots, if the mould was not the leaft damaged, which is feldom the cafe.

THE white infects on pine-plants are very pernicious, and have been the ruin of many plants that would have produced good fruit if they had not been infected; for when they are full of them, the fruit

K 4

never

never has that good flavour as when the plants are clean. Befides the damage done to the fruit, they greatly retard the growth of the plants, and make them very unfightly.

THOSE that are infected, if not kept under by cleaning, will be totally deftroyed, and cleaning them by hand is attended with much trouble, and at a confiderable expence; to brufh and clean them properly will take up a great deal of time; befides in doing it the leaves are often broke and the fides fcratched, fo that the leaves decay in patches, which makes the plants very unfightly: and what is worfe, after all that trouble and expence, they are as bad in a few weeks as they were before; yet, if they were not cleaned when moved, they would grow fo dirty that the infects would quite deftroy the plants.

MANY things have been tried, and even advertized as effectual remedies to deftroy

deftroy them; and all of which have been
fo pofitively afferted as infallible, that it
may look like an impofition on the public
to offer any other remedy; but in fupport
of what I here advance, and for the fatif-
faction of the public, I fhall give an account of the effects of a fure and fafe remedy, having from repeated trials found it
effectual, otherwife I fhould not prefume
to publifh it.

For many years, when I lived with
William Salvin, Efq; of Croxdell, the
pine-plants were every year in a moft
dirty condition, and every time they were
to move or fhift, a deal of time was fpent
in cleaning and brufhing them; yet notwithftanding all the pains that was taken
with them they every year grew worfe:
the bottoms of the fruiting-plants, when
they were moved in the fpring to ftir the
bark and add a little new to refrefh the
heat, were as white as if they had been
firft wetted and then dufted with meal;

and

and amongſt the white ſtuff were many of
the white creeping vermin, ſome of them
very large, ſome ſmall, and all the leaves
full of white ſpecks.

I TRIED every thing I could imagine,
but to no purpoſe; and every thing that
was advertized, but without effect; therefore I deſpaired of ever finding a remedy:
at laſt, however, I had the good fortune
to ſucceed, and in one ſummer had not a
ſingle ſpeck on them in one houſe, and
greatly leſſened in the other that was ſo
very dirty. On theſe I could not perform the operation that ſpring completely;
but it was done in the autumn, and they
were very ſoon both clean, and remained
ſo. This was ſix years before I left the
place, which was ten years ago, and there
was no occaſion for any repetition of the
remedy.

I TRIED its efficacy on ſeveral hot-
houſes which I had built for gentlemen,
and it always ſucceeded. Some of my inti-
mate

mate acquaintance, to whom I communicated the secret, for then I had no thoughts of making it public, have all tried it, and found it answer. Some of them, at a distance, have informed me so by letter.

I MADE an experiment at Sir Thomas Gascoigne's, Bart. when I came to the place, May-day, 1771. The fruiting-plants were just shifted into fresh heat, and the younger plants into larger pots: they had been cleaned before they were plunged, but there were many insects all round the bottom leaves. I applied the remedy; they went off insensibly, and there have been none in Sir Thomas's hot-houses since.

I ALSO tried another experiment, which is a convincing proof that none of those insects will live where this remedy is used. Sir Thomas had a quantity of plants (in 1772) from abroad, crowns and suckers, which were very full of vermin. I applied the remedy without cleaning, and planted them in the middle of the hothouse

houſe amongſt clean plants : they dropped off the infected plants, and did no
damage in the houſe, which is a certain
proof they were all killed.

SIR THOMAS's ſtoyes being all new,
there was no place for the vermin to lodge
in, neither had the infection been violent,
and of ſo ſhort a duration, as two years,
that the vermin were in the pots and on
the plants. The houſe was not ſmoked,
neither was the tan ſifted, only the mixture applied, which I did by way of preventing the ſpreading of the inſects, with
an intent to go through the operation in
the autumn ; but as I found, long before
that time, they were all deſtroyed, I proceeded no further, and have never ſeen
one ſince.

THE white ſpeck on the leaves of pines
is the ſpawn of the white creeping inſect,
and is depoſited on the leaves of the pines
much in the ſame manner as the caterpil-
lars are on the leaves of cabbages. I have
viewed

viewed them in a good microfcope, and found all the parts of the vermin complete under the white fcale, when taken off at a proper age without being bruifed, and when they come to a certain period they force the fcale from the leaf, and fo defcend to the bottom to grow to maturity.

MANY have thought the white infects on the leaves to be inactive and not capable of moving; but that is impoffible: they have no life at their firft appearance on the leaves; for if brufhed off foon they leave no mark on the leaves, are extremely thin, and quite dry; but when they have remained fome time on them, they become thick in the middle, and if brufhed off at that age, a foft gluey matter comes from them, which is the bruifed infect arrived to a fubftance; but the fkin is fo thin and tender that it is broken by a flight touch.

THAT they have life is evident; for where they ftick to the leaves, although
brufhed

bruſhed off before they come to maturity, the outer rind of the leaves is eaten through, and the place becomes quite white and dry. That they fall off when come to ſuch an age is certain; for many of thoſe white ſpots may be ſeen on infected plants that have not been cleaned.

WHAT makes me ſo very particular in this is, that many people imagine the white ſpeck to be an inſect of itſelf, and the white creeping vermin to be different; but it is no more than the parent of the other that does all the miſchief, and therefore muſt be deſtroyed. As a proof that a white ſpeck is the ſpawn, there is often to be ſeen on infected plants a ſmall white ſcale on the very end of a leaf, and not another near it. No inſect can breed without ſeed, and there muſt be ſomething to form that white ſpeck. They are very ſmall at firſt appearance, and grow gradually to a certain ſize.

THE white creeping vermin may with
pro-

propriety be called the pine-bug; for it lodges amongft the bottom leaves of the plants, and in the leaft hole or chink in the houfe. If the ftove is old, and has been infected fome years, there are thoufands of them, and their fpawn lodges in the crevices and amongft the tan; all of which muft be deftroyed.

WHERE there is only one ftove, the latter end of Auguft or the beginning of September is the only time that the operations for a thorough cleaning can be performed, as there is no making a perfect cleaning of plants in fruit.

THOSE that have breeding and fruiting-ftoves feparated, fhould begin with the young plants the beginning of March, and then the plants will not be retarded in their growth. The crowns may be put, after they are dreffed, into the houfe that is cleaned; and the fuckers, as foon as they are taken off, then there will be time and opportunity fufficient to get the

fruiting-

fruiting-houſe perfectly cleaned, which ſhould be done as ſoon as poſſible, in order that the fruiting-plants may be put into their fruiting-pots in good time.

ALL the young plants muſt be removed out of the houſe to ſome convenient place where they can remain with ſafety for near twenty days, for ſo long will it take to get the houſe completely ready for them. The weather at that ſeaſon being ſometimes very cold and hard froſts, they muſt be in a place where they can be protected from the inclemency of the weather, and have ſome heat. The flue of the fruiting-ſtove will be the propereſt place ; and although attended with ſome little inconvenience, it will be only for a ſhort time.

THEY ſhould have a board ſet under them, which muſt have ſome bricks under it, to make a vacuum between the flues and the plants ; for if they are ſet on the bare flues it will dry their roots too much and cauſe them to fruit. They

3

muſt

muſt have a little water every day while they ſtand there.

WHEN the plants are removed, throw out all the tan, and remove it to ſome diſtance from the ſtove ; then make the houſe as clean as it can be made with a bruſh and broom ; after which provide three or four chafing-diſhes, into which put ſome red-hot cinders, ſet them in the bottom of the pit in the ſtove, and throw into each an ounce of rock-brimſtone broke into ſmall pieces ; then ſhut up the houſe cloſe, and let it remain ſo until next morning. The ſame operation muſt be performed for three days, which will effectually kill all the vermin and their ſpawn that are lodged in the crevices of the houſe. If there are no chafing-diſhes, garden-pots will do as well.

As the ſtove is empty, it would be right to point the inſide, and point and ſtop up all the crevices in the plaiſter ; but before that is done, all the inſide of the houſe

VOL. II.   L   ſhould

should be washed with a sponge dipped in vinegar, which will take off all the steam left on the rafters and glass by the smoke of the brimstone.

AFTER all the wood is painted, and all the cracks of the plaister pointed and white-washed, give the house a good deal of air night and day to dry the paint and white-washing. There should be nothing put into the stove for six or seven days, by which time the paint will be quite dry, the sulphureous smell gone, and no ver-min of any kind left alive in the stove.

THE tan that was carried out should be all sifted through a coarse riddle, the rough put into the stove, a layer of new bark six inches thick at the bottom, then a foot of the riddle bark, and continued until the pit is full, leaving an open trench at one end ; and trench the whole bed over, mixing it well; then lay it level, and let it re-main so to heat. In about a week or ten

days

days it will come to its full heat, and be ready to receive the plants.

THE reason for sifting the bark is, because there are many of the live vermin amongst it that will go through the riddle with the small tan, which should not be used amongst any mould that is for the stove, or where there is any heat, those that are alive will soon be killed by the cold; but as there will be much of their spawn amongst it, if used where there is heat they will soon come to life, and be as troublesome as ever; therefore it should be carried and thrown upon grass-land, for which it is very good if the land is of a clay or stiff nature; but is very pernicious if of a gravel or sandy kind. If there is no such use for it, it may be thrown into the fold where cattle are fed, where, being mixed with the long straw, it will make good dung.

THE ingredients proper for killing the white insects on pine-plants.———Four

L 2   pounds

pounds of flour of brimſtone, one pound of Scotch ſnuff, two ounces of the leaves of walnut dried and ground to powder, and finely ſifted ; this is the proportion.

THE quantity muſt be according to the number of plants that are to be dreſſed. Let them be well mixed ; then take the plants out of the pots, cut off all their roots cloſe to the ſtem, which examine carefully, for ſometimes the vermin eat holes into it, and many of them will lodge therein.

IF there are any holes, pick out the ver- min with ſome ſharp-pointed thing, and fill the holes with the mixture, for there is often ſpawn in them as well as ver- min.

TAKE ſome of the mixture in the hand and rub the ſtem well, then drop ſome of it between every leaf; take ſome of the mould that was made up for the crowns and ſuckers at their firſt potting, after

taking

taking them out of the tan, and to four barrows of the earth add one pound of the mixture: let the earth be well mixed with it, and then fill halfpenny and penny pots, according to the largeneſs of the plants, and plant them, taking care to keep the plants upright after the mixture is put in between the leaves, that it may not be ſcattered. They then ſhould be plunged into the bed according to its heat, either half way or to the top; they will bear more heat than thoſe plants that have roots, and they require it to make them ſtrike.

THEY ſhould have no water for ten days after they are plunged; then they ſhould have a little, for by that time they will begin to have ſome roots, and the water muſt be poured ſlowly into the hearts of the plants, that it may not force out the mixture from amongſt the leaves, but car-ry it down cloſer to the bottoms.

AFTER the water has been ſo poured on them for five or ſix times, it ſhould be given

L 3

more

more freely, to wash out the mixture from the heart of the plant, which must be in about twenty days after being plunged: from this time there should be no more water poured into the heart, but in amongst the bottom leaves. This mixture will no ways hurt the leaves, nor retard the growth of the plants.

As soon as the roots will keep the ball together, they must be shifted into larger pots to bring them to a regular course; but there will be no occasion to use any more of the mixture neither in the mould nor to the plant, for the vermin will be effectually destroyed, and never give any further trouble.

WHEN the plants are to be plunged, the bark should be stirred no deeper than just to receive the pots; for as it has been so lately stirred and mixed from the bottom, if it was now stirred any depth the heat would be greater than the plants could bear.

IF

IF the plants that are in fruit are very dirty and full of vermin when they are removed in the fpring, in order to add frefh heat after the dead leaves are pulled off and the mould in the pot ftirred at top, fpread a little of the powder over the top of the pot, and put fome in between the leaves, which will preferve the fruit-ftem from vermin, and confequently the fruit and crown; fo that the fruit will be clean to go to table, and many of the vermin will be killed, but the houfe will not be quite clear; for the live vermin have many recefles amongft the roots and bark where the infection is great, and has been of a long ftanding.

As foon as the fruit is cut, take off the fuckers, and rub the lower part of the ftem well with the powder, putting it in amongft the fmall leaves at the bottom, alfo between every leaf, and into the heart; for there many of the live vermin often lodge: they then may be put into the bark in the clean ftove very fafely to ftrike root.

WHEN the plants are dirty, the under leaves of the crowns are generally very dirty alfo. When they are taken off the fruit, fome of the powder fhould be rubbed on the under fides of the leaves, and dufted between every leaf, heart and all, and may then be put into the bark in the clean houfe.

WHEN the fruiting-houfe is cleared of the fruit, it muft be treated in every refpect the fame as has been directed for the breeding-ftove, and then I am in hopes there will be an end of thofe troublefome vermin, which have been a great plague to Pine-plants for many years.

IF any plants, crowns, or fuckers come from abroad, whether any vermin appear on them or not, it would be right to duft them before they are put into the ftove. If any fingle plant in the ftove fhould by any accident efcape being cleaned, lay fome of the mixture all over the pot and between the leaves, and the infects will foon follow their companions.

WHERE there is but one stove there is an unavoidable necessity of retarding the next year's fruit, for there is no doing any thing then with the plants in fruit, but what is before-mentioned, and that is only a superficial operation. It will not be sufficient to clean plants that are much infected, for there is no meddling with the roots of plants that are in fruit, neither is there a possibility of fumigating the stove or riddling the bark, for it would keep the plants in fruit too long out of heat, and stunt them.

IF there is no partition between the fruiting-plants and the young ones for next year, the young ones should not be shifted after May, but have plenty of air, water, and a moderate heat, to keep them growing all summer until the fruit is all cut, for then they must have all their roots cut off.

As the fruit should be forwarded as much as possible and cleared, it will be
better

better to lose a few infected fruit than be
too late in getting the plants for next year
cleaned and put in order, as they have all
fresh roots to make; and there is no doing
things effectually if any infected plants are
in or near the stove.

THE case is very different in having
plants that never have been dressed, which
are much infected, in pots near clean
plants, and plunging a dressed plant that is
full of vermin amongst them; for the live
vermin, whose spawn dirties the leaves of
the plants, lodge in the pots amongst the
roots and in the tan, and there is no possibility of being certain of coming at them
with the powder. Those in the pots may
escape into the tan and soon increase their
numbers, so that all the labour that has
been bestowed will be lost.

WHEN a plant is taken out of the pot,
and all its roots cut off, there are none of
the vermin that are come to life but what
will be destroyed; for if any of them are
lurking

lurking amongft the bottoms of the leaves, they are fure to be killed as foon as the powder touches them, which is eafy to accomplifh, as there is no fhelter for them; but in pots of infected plants they have many receffes, and may fecrete themfelves fo that none of the powder can touch them.

THE powder is fo ftrong a poifon for that infect, that it foon kills it; and I am of opinion that in two years the moft infected houfe might be cleaned by fifting the tan once, and ufing the powder at every fhifting, and putting fome between the leaves; but this is only conjecture. Although I tried it in the fpring moving, and found them greatly diminifhed, I had not patience to go through the experiment, as I had got one houfe clear by the firft-mentioned operation.

IF any perfon choofes to try the remedy in this manner, there fhould be the fame quantity of the mixture in the

mould

mould as before directed; and when the
plant is taken out of the pot, and the
roots dreſſed, the ball ſhould be duſted all
round, and then filled up with the mould,
and the ſame quantity put between the
leaves.

THIS, in all probability, would kill all
the vermin that are in the pots at the time
of the operation; but if the houſe is much
infected there will be great quantities;
and when the powder, by length of time
and watering, may have loſt its ſtrong
poiſonous quality, there will be many in
different parts of the houſe which may in-
creaſe and come into the pots.

THIS is only what ſeems probable to
me, and would be a great advantage to
thoſe that have only one houſe, that the
plants might then be ſhifted at their
proper time; but as I cannot aſſure a
complete cure, only opinion, I ſhall proceed to give ſuch directions as will be
effectual.

IF

IF there are a few late fruit in the houfe, it will not be worth while to wait for their ripening; a bed may be made for them in the Melon ground, which will be very eafy. If there are any pits for Melons, fill one end the length of a four-light frame, which will hold a good many plants, with old bark from the ftove, and if it is cold add a little new; mix them well, and fet the frame on the top of the bark.

IF the depth of the frame is not fuffi-cient for the height of the plants, raife the frame with bricks, and clofe up the vacancy with tan on the outfide, which may lie floping to the fide of the pit. Plunge the plants in the bark, cover them with glaffes, and they will ripen there as well as in the ftove. By being covered with mats at night they may be kept to the middle of November, by which time all that are good will be ripe.

THE fooner this is done the better; for
if

if a frame or two will hold all the crop
that is left, it would be worth while to
clear the houſe by the middle of July (as
there is no new erection nor expence,
only a little labour) then the next year's
plants would be very little retarded.

THE bark the plants are plunged in
will be very uſeful. As ſoon as all the
fruit is cut let it be ſifted, lay the round
under cover, mix it with half as much
new, and turn it once or twice, and it
will be much better than all new to re-
freſh the bed the next time the plants
are potted, and to fill up; for by being
turned from the bottom it will be ſunk a
good deal below the level of the pit.

WHEN the houſe and bed are prepared
as directed for the young plants in March,
and the heat comes up, the plants muſt be
all taken out of the pots. The young
plants muſt be treated in the ſame
manner that the plants were in the
ſpring.

THE

THE large ones that are for fruit next year fhould have all the mould fhaken clean off, and all the fmall roots cut clofe to the ftem, and only a few of the fineft white fhort roots left on, that they may ftand an inch or two within the fides of the pot.

EXAMINE and rub the ftem as before, and duft all the roots that are left; but they muft not be rubbed, for it would break all the young fibres and caufe the roots to rot; yet it is neceffary they fhould be dufted, as there is a probability of their being infected with the fpawn; then dafh with the hand, between every leaf, as much of the powder as a fmall nut-fhell would hold, and keep the plant upright after the powder is put into it.

THE mould directed for plants whofe roots were fo much burnt as to require fhifting, is very proper for both old and young plants at this potting.

To

To four barrows of mould add one pound of the powder, and let it be well mixed. Three-halfpenny pots will be a proper fize for the fruiting-plants, a little mould being firft put into them. The plant fhould be held up by the hand, and the mould put in amongft the roots, fo that there may be mould between them, and none of them clofe together. They fhould all lay eafy, and not be twifted, and kept an inch or two from the fides of the pots. The mould fhould be faftened all round the fides of the pot, the plant made faft at the neck, and plunged as foon as poffible.

The day after the large plants are plunged they fhould have a little water, which fhould not be poured amongft the bottom leaves, but all over the tops of the pots: they will require no more for ten days, by which time they will begin to have fome young roots, fo that they muft be refrefhed with water according to the heat of the weather, and the water

fhould

should be poured in amongst the bottom leaves.

THEY should have a great deal of air until they have got fresh roots; for if the days are hot, and they have little air, their leaves will draw up, and many of them will decay. Those who have covers for their stoves would do well to cover them all the middle of the day; and if it is fine weather, after they have got roots and want water, it will be of great service to them to be sprinkled all over once or twice a-week.

THE young plants should be planted in penny and halfpenny pots, according to their size, and plunged into the bed; but should have no water until they begin to have some fresh roots, which will not be so soon as the large plants, for the large plants being in larger pots and deeper in the bark, and having a much longer length of stem to push, they strike root sooner

VOL. II.   M   than

than the small plants, whose stems are short and pots little.

I T will be fourteen or fifteen days before they want any water, and then only a little, which must be poured in amongst the lower leaves, because there are no roots but what are close to the stems, and they must be managed in every respect the same as the plants that were dressed in the spring.

As soon as the young roots will hold the ball together, they must be shifted into larger pots; the fruiting-plants into the pots they are to fruit in, and the succession-plants into three-halfpenny pots. Great care must be taken not to break the ball, for that would greatly retard their growth; for as the plants are shifted before the roots mat, or any small woolly stuff grows, there is no occasion to disturb the roots to break their small fibres, which will grow as soon as they are shifted into larger pots.

THE

THE large plants that are planted in their fruiting-pots fhould have the fame mould as thofe that were before cleaned and fhifted at the proper feafon; but the young plants fhould have that which was directed for plants at the fecond fhifting; and if they meet with no misfortune, their balls need not be reduced in fpring, but only removed into larger pots.

IF they are managed as directed, the pots of the fruiting-plants will be full of roots in a little time after they are fhifted, and produce good fruit, but will be later by a month or fix weeks.

WHERE there is only one hot-houfe, the greateft difficulty is the keeping the plants while the ftove is cleaning and getting ready, for which reafon it fhould be done as foon as poffible; for if the weather is very cold and frofty it will be very difficult. If there are but few fruit, then it is not worth making a bed for

M 2   them;

them ; it would be better to throw them away than to be too long in clearing the ſtove.

IF the houſe is cleared at the latter end of July or Auguſt, the ſhed of the ſtove (as all ſtoves have or ought to have one) will anſwer very well. They muſt be left in the pots not crouded, and have air every day, but no water ; for if any water be given them, they having no bottom heat, it would kill all the roots, which would be of great diſadvantage, as then there would be no freſh roots to leave on the fruiting-plants ; it would alſo turn the plants very yellow, and rot the hearts of the young plants, which are much ten-derer than the large ones.

I AM convinced that ſtoves which have been only a few years built, and not much infected, may be cleared by daſhing ſome of the powder amongſt the leaves, and lying a little all over the tops of the pots.

I have

I have cleared feveral by that method ; and by following the whole of the directions the moft infected hot-houfe will be as clean as if there never had been any vermin in it.

M 3   CHAP.

C H A P. XIV.

Of Mushrooms.

TO pretend to raise plants without
seed is a contradiction to nature;
but what is here meant in the raising of
Mushrooms, is to produce them on beds
made for that purpose, without spawn
being planted in them, as is the common
method of propagating them.

MUSHROOMS are wanted in every family, and in most places where there is
any thing of a garden there are beds for
raising them.   In some places they succeed
very well for a year or two, and then there
is often a loss for a season, sometimes
longer.

THE getting of good spawn is frequently attended with great difficulty;
                                                    and

and when it is got, if the bed is not in proper order to receive it, there is a very great difappointment, which may and often happens to thofe who are well acquainted with the different methods which are now in practice for propagating them.

THERE is no plant that is raifed by art, where nature has given fo many hints to facilitate its propagation, as the Mufhroom; and yet it is ftill in its infancy. There is not in common practice (that I know of) any certain method to raife them fo as feldom to fail of producing a good crop.

IF the beginning of the fummer is warm and dry, and afterwards there fall kindly warm fhowers, the fields produce them in plenty; whereas if the beginning of the fummer is cold and wet, and afterwards becomes warm and dry, there are few or none.

M 4 ON

ON hot-beds, where cucumbers and melons have been planted, and have not been fuccefsful, there is often a quantity of mufhrooms.

IN large dunghills that are laid in the fields, where there is a good deal of long ftraw, which has not laid wet in the fold, in a dry warm feafon there will be much fpawn and fome mufhrooms.

IN ftable-yards, where a good many faddle-horfes are kept in the ftables, whofe litter is thrown out long, and a large quantity every day, if the feafon is dry and no wet lies in the place where the dung is, the bottom of the dung will be full of fpawn in the autumn.

IF a quantity of litter, that is juft moift, (four or five loads) be mixed with light loam or fand, and thrown in a heap in a dry place, it will produce fpawn in three months.

ALL

ALL thefe examples make it evident
that mufhrooms may be produced without
fpawn being planted, and that they grow
in places where there never were any before, as all muft allow.

BY what means this plant (if it is proper to call it a plant) grows in fuch places
I do not pretend to demonftrate; it is certain no plants grow without feed: there
is no new creation.

THE feeds of mufhrooms being fo very
light, it is not improbable but that in the
fields where mufhrooms grow it is carried
over the whole ground, and lodged in
every little hole and crevice, where it grows
as foon as it meets with proper materials.
The reafon why mufhrooms do not grow
in all the places where fuch ftuff lies is,
becaufe it is not in proper order, being too
wet or too dry, or the feafon fo cold that
the feeds cannot vegetate.

IT may have been imagined long before
this

this, confidering the many and great im-
provements in the propagation of all kinds
of plants, that mufhrooms would not have
been forgotten ; and thofe hints that nature has given would have induced fome of
the many ingenious to attempt bringing
them to perfection without fpawn.

I t was from thefe obfervations that I
was encouraged to make fome trials ; and
although I did not fucceed at firft fo as
to produce a good crop, yet I found fomething might be accomplifhed, and by per-
feverance I difcovered a compofition that
produced finer mufhrooms, more in quantity, and the beds continue much longer
in bearing than thofe planted with fpawn
in the common way.

I have ufed no fpawn for many years,
and have never failed of a good crop. I
fhall give a full account of the method I
have followed, and if the directions are
obferved accurately, can affert it to be a
fure way to bring them to greater perfection

tion in quality and quantity than any other artificial method I have feen practifed; they are fuperior in thicknefs to thofe that grow naturally in the fields.

THE proper times for making the beds are as follow: to have them in winter, the beds fhould be made the beginning of July; for a fpring crop, the beginning of December; and to have plenty all the fummer there fhould be another bed made the beginning of March. Thefe three beds will produce them every month in the year.

IF the beds are made according to this new method, they laft much longer than thofe planted with fpawn.

BEFORE we proceed to give directions for making the beds, it will be proper to fay fomething of what are the proper places for them.

THOSE that are very fond of mufhrooms in winter, may be certain of having them

3   in

in great plenty if they will be at the trifling expence of building a ſhed againſt
a ſouth-wall, with a pit three feet deep
and four feet broad, the length according
to the quantity that is wanted.   Thirty
feet long will produce ſufficient for a large
family.

THERE muſt be a flue all round the pit,
two feet deep and eight inches wide, and
a vacuum of an inch on the inſide next
the bed, which ſhould be left open at top
that the heat may come up to warm the
top of the bed, which it would not if the
vacuum was cloſe.

THE vacuum all round prevents the
heat from drying the ſides of the bed ; for
if the flue had no vacuum, the heat would
deſtroy all production for a foot round at
leaſt ; and when the bed is covered with
ſtraw and litter (as it muſt all winter) to
keep off the froſt and cold air, the ſtraw
ſhould be laid all over the flue and vacuum, which will cauſe a gentle heat all over

6   the

the top of the bed, and greatly encourage the growth of the muſhrooms.

There will be no danger of the ſtraw firing, as very little heat will be ſufficient to keep the muſhrooms growing.

Where there is a hot-houſe for pines the ſhed will anſwer pretty well for making a muſhroom-bed for the winter and all the ſeaſon. As the fires are kept on they will grow very well; but it is too cold for the ſummer's production, for they love the glaſſy heat of the ſun, or an artificial one, although they cannot ſtand its bright rays.

In winter they will grow in dry cellars and produce a good crop; they will alſo thrive extremely well in the end of a ſtable where there are a good many horſes kept. They will thrive and produce a great crop in a ſhed facing the ſouth, from the beginning of April to the end of September; but will produce nothing in winter

ter, as the air is too cold for them, although the bed be covered ever fo well with ftraw and litter.

THE materials and method of making the beds, either for winter or fummer, or which of the above-mentioned places they are made in, being the fame, I fhall only give directions for making that in the fluid pit for winter ufe.

THE middle of June provide fome frefh litter from the ftables ; a bed thirty feet long and four feet broad will take ten cart-loads : it muft not be what has laid long and is turned black ; but if mouldy and long will anfwer very well to mix up with fome frefh, at the rate of three parts frefh to one of mouldy. Throw it up in a heap under cover, for it muft get no wet, and let it remain for eight days, by which time it will have a gentle warmth, and it is then fit for ufe. Get fome frefh tanner's bark ; a thirty-feet bed will take three large cart-loads.

It

IT fhould be laid in a heap under cover ten days at leaft, for bark is much longer in heating than dung; when it has been in the heap fix days it fhould be turned over, that the outfides may be in the fame condition as the middle; for the outfide of bark never heats when it lays in an open heap; when it has juft come to a moderate heat after turning, it is in order for the bed.

OLD bark from a pine-ftove that is dry and round will anfwer, if it is watered a little and thrown into a heap to ferment; but it is not fo good as the new; it was from fuch I had the firft fuccefs. I then mixed it with an equal quantity of new, and it anfwered much better than when it was all old; but there being a difadvan-tage in taking fo much round bark from the ftove, I tried all new, and found it much better than either of the former.

AT the fame time there fhould every day be faved from the ftable the horfe-drop-pings,

pings, which fhould be kept as whole, dry, and clean from ftraw and hay as poffible. They fhould be laid under cover, and fpread two inches thick. The place where they are laid fhould be airy and dry, for the drier they are when laid on the bed the better. If they are got before any of the other materials, and kept dry and not moved, they will be no worfe.

To make the bed put a layer of the dung that was fhaken up, two feet thick, the breadth and length of the bed; it muft be made very equal, and pretty firm, but not trod; then a layer of bark, four inches thick; over that a layer of the fame dung, a foot thick; and on that a layer of bark, two inches thick; and on that fix inches of the fhorteft of the dung that was thrown up, which finifhes for the prefent.

WHEN the bed has been made ten or fifteen days it will then have a gentle heat; lay another layer of bark, two

inches

inches thick ; and on the top of that a layer of the horſe-droppings, three inches thick (they ſhould be taken up carefully and as little broken as poſſible) ; and on the top of the droppings an inch of the ſhorteſt of the dung from the heap that was ſhaken up. This finiſhes the bed.

THE bed ſhould remain ſo for ſix or eight weeks, by which time it will have a moderate heat, if all has ſucceeded as it ought, and will be fit to lay the mould on. It ſhould be juſt milk warm. The mould ſhould be laid on three inches thick, but muſt have no water, until, upon examination of the mould, there appear little white ſtrings amongſt it, which will be in about three weeks after.

IT muſt then have a little water once a-week, and if it was made milk-warm, or taken out of the ciſtern in the hot-houſe, it would be much better. In about three or four weeks after the white ſtrings ap-pear, the muſhrooms will begin to come ;

the bed fhould then be refrefhed with water often, but not much at a time; for if there is much given at once it will kill all the mufhrooms that are above ground, and greatly damage the bed.

A BED thus made will continue to bear very plentifully for half a year, fometimes a whole year, and produce much finer and larger mufhrooms than any of thofe that are produced in the fields.

THE mould proper to lay on the bed is two parts of very rotten dung fifted, two parts of fine light loam, and one part of rich black kitchen-garden mould broken fine, and all well mixed.

IF a winter-bed, it muft be conftantly kept covered with dry litter three inches thick; firft fhaking all the dung out of it; and then three inches of clean white ftraw, and a gentle fire kept when frofty and cold weather, or when there has been long dull rainy weather, which is

very

very prejudicial to them; for they rot as they come out of the mould.

In such weather it would be very proper to take off the litter, and lay on the ftraw next the bed, as by that means it will admit a free air and prevent mouldinefs; but in frofty weather the litter fhould be next the bed.

If the winter-bed is made in the fhed of the pine-ftove, it fhould be made at the fame time and manner as the other, covered with the fame fort of mould, and managed in every refpeét the fame. But as there will be two fires in the fhed all winter, which will keep it warm, there will be no occafion to lay any litter, only a very thin covering of ftraw.

The bed muft be made againft the north wall of the fhed, for the ftove-wall would be too hot. The forefide of the bed will dry much fafter than that againft the wall, and muft have a little water oftener, for

N 2   a very

a very little will ferve the backfide, as a
great dampnefs comes through the wall,
efpecially if moft of the fhed is under
ground, as it generally is.

THOSE that have no fhed with a pit, nor
a ftove-fhed to make a winter-bed in, may
make them in a dry cellar, at the infide
end of a warm ftable, or in the corner of
any outhoufe where there is a conftant
fire.

THERE will be no occafion for any cover
in a cellar in winter, as there is no cold
air; and in a ftable the air is both damp
and warm : they will require little water
in a cellar, and hardly any in a ftable. If
in a houfe where there is a conftant fire,
they will require a little water but no
cover. In all the places where they are
made, they are all of the fame materials.

THOSE that choofe to have mufhrooms
in the fpring and fummer only, fhould
make a bed the beginning of December,
and

and it will come into bearing the end of March or the beginning of April, and will produce great plenty all spring and summer.

They should be under cover facing the sun, but never exposed to its rays, nor to rain; for it is impossible to manage a mushroom-bed right where it is exposed to rain: no covering of mats or straw is sufficient to protect it.

Suppose the bed had been watered in the morning, and in the afternoon a strong thunder shower should fall, it would go through all the covering, and perhaps destroy the bed entirely. Many a good bed in full bearing has been so spoiled by such an accident, that it never produced a single mushroom afterwards.

If a shed with a pit thirty feet long has the fire-place in the middle on the backside, and made to draw both ways, and to come up the middle to divide the pit

N 3

in

in two, and one of the ends made every June, there would be fufficient for a middling family all the year.

WHEN an old bed is deftroyed, if it has worked kindly and been managed well, the tan that was laid above the firft layer of dung will be very dry, white, and mouldy, and fhould be faved, it being exceeding good for laying two inches thick on the new bed, below the three-inch layer of horfe-droppings : it will caufe the bed to bear three or four weeks fooner, and the mufhroons to be much larger and thicker in the flefh ; for the whole top of the bed will become one folid mafs of fpawn.

WHEN the old beds are pulled to pieces there will be much fine fpawn in them ; but let not that tempt any one to lay it on the new bed, for it will entirely fpoil it. Although by fuch means you might get mufhrooms rather fooner, they would be fmall and bad, for the waterings that

would

would be neceſſary to bring theſe bad muſhrooms forward would prevent the production of any new ſpawn.

BEDS made in cellars or houſes where fires are kept, and in ſtables, will not produce muſhrooms all ſummer; for as the weather grows warm the air in the cellar grows cold, and the beds in the ſtables will be too moiſt if horſes are kept there all ſummer, and too cold if there are none; ſo that after April there will be few muſhrooms on any of them.

MUSHROOM-BEDS made after the manner here directed are far ſuperior to thoſe made in the common way; they produce much finer muſhrooms, laſt much longer in bearing, and have greater quantities on them at a time.

I HAVE gathered great quantities of muſhrooms from beds made according to this method, which were much whiter than thoſe gathered in the fields, a great

N 4   deal

deal firmer, and when they were as cloſe as the ſmalleſt buttons that grow natural, many of them were four and five ounces weight.

IF all be managed as directed, you will be ſure of ſucceeding; but if any one ſhould fail, let him not deſpond at the firſt trial, there being a poſſibility of erring, though no great probability, for I have not omitted the leaſt article in the prepara-tion of the materials, making and manag-ing the beds: but information from books written with the greateſt candour may be difficult, until a little practice renders it eaſy.

CHAP.

C H A P. XV.

On Asparagus.

THE common method of managing Asparagus beds is so well known, that it is not necessary to say any thing on that head; but there are several places in England that are remarkable for having larger asparagus than the common run of the country, as Battersea and Gravesend, near London, which some may attribute to a different management, which is not the case, but the goodness and depth of the soil.

I THINK that Pontefract, in Yorkshire, is greatly preferable to the above places in the goodness of its soil, and might produce larger asparagus.

THE

THE largeneſs is the thing ſo much talked of at the above places, for I never heard that the beſt judges ever aſſerted that it was any ways ſuperior in taſte to thoſe raiſed in other parts of the kingdom.

THE ground in the above places is not all of the ſame kind; that at Batterſea is a loam of a great depth; the ſoil at Graveſend is very light and ſandy, and alſo very deep.

THE ground at Pontefract is richer, deeper, and finer than either of the above; it alſo inclines to a brown ſand, is the beſt ſoil I have ever ſeen, and I am confident, if the aſparagus beds were managed as they are near London, they would produce much larger than any in the kingdom in the common way.

I HAVE ſeen and taken notice of the dreſſing of the aſparagus beds at Batterſea and Graveſend, but cannot ſay there is any material difference in their method
from

from the general run ufed about London, only they lay a deeper cover of earth than is commonly practifed in moft places, the reafon for which is, that their foil is lighter.

As I fee fo little difference in the method of the culture of afparagus in thofe two places, from that of their neighbours, I can attribute their fuperior fize to nothing but the fuperior goodnefs of their foil.

It is evident that the foils of the abovementioned places, notwithftanding their difference, are all very good, and produce fine afparagus, which is a plain demonftration that there are different foils, that the fame fort of foils may be planted in, and that they will thrive equally well in them all, which is a thing that fhould be paid more regard to than generally is the cafe.

It is a general error in writers who
treat

treat on these subjects, to mention only the soil that such and such plants will grow the best in, and that in such a manner as to leave their readers to think that they will not grow at all in any other kind of soil, which often discourages people from sowing or planting many different kinds of roots, shrubs, and trees, in soils where, though they do not attain to the greatest perfection, they may become very useful, profitable and ornamental, and some-times to as great perfection as they do in that soil which was thought peculiar to them.

Asparagus will thrive in all light, rich, deep soils, and all sandy soils may be made rich with rotten dung, so as to produce very good asparagus; and in every place it is planted in, if it is not light and rich by nature, it must be made so by art; so that there is no place in England but where good asparagus may be had, and at no great expence, as there are few places but where the proper materials may be had

had to make fuch a foil, which is what I
fhall give directions for in the common as
well as the new method, for growing it
to a very large fize.

As the largenefs of afparagus is what
is greatly admired, it may be brought to
grow to a very large fize in every gentleman's garden ; but many will object, efpecially thofe who have fmall kitchen-gardens, and have little ground to fpare, that
it takes a much greater fpace than in the
common method ; yet I do not know, if
the product was to be weighed, but that
the largenefs would make up for number,
though I have not as yet tried that experiment.

In any part of the garden where no wet
ftands in the bottom (for where water
ftands afparagus will never thrive, as it
kills all the roots in winter, they ftriking
down to a great depth) make a trench two
feet broad at three feet diftance from the
border ; three feet from that another ; and

fo

fo on for as many rows of afparagus as are intended to be fown.

Dig and carry all the earth out of the two-feet trench, which may be difpofed of in different parts of the garden where the ground is low. The trench fhould be two feet and a half deep at leaft.

If the foil in the garden will not allow fuch a depth, fome of the earth taken out of the trench may be laid on the three feet between the trench to heighten the ground, for it will not be very material though the ground be a foot higher than the other parts of the garden; but the bottom of the trench muft not be dug lower than the other parts of the bottom.

Lay in the bottom of every trench three inches of good rotten dung, to which add three inches of the natural mould, if it is light; but if of a ftiff nature, none of it fhould be put in. If there is any mould

added,

added, mix the dung and mould well by pricking it over three or four times with a spade.

THE proper time to do this work is the beginning of February, that the mould which is laid in the trench may be well settled before the middle of March, which is the proper time to sow the Asparagus.

THE summer before this is intended to be done, the composition for filling the trenches should be prepared, and turned several times, that it may be well mixed and incorporated before it is used.

THE materials of which I have made the composition, with the greatest success, are as follow : Sludge from a river that runs slow and is muddy; but the best is where the tides flow and ebb ; though if neither of these can be got, the sludge from the bottom of an old pond that has not been cleaned for a long time will answer very well.

LIGHT

LIGHT peat-mould that lays at top and is very foft; if it is difficult to get, as it is in many places, the bottom of wood-ftacks, or the places in woods where much leaves and fticks have rotted, the light mould to be got there will do. Light rich loam from a common pafture. Fine rotten dung that has laid two years, and has been turned feveral times, and reduced to a fine mould. Sharp fand from a pit, or river-fide, which is better. Thefe are the ingredients.

THE quantities of each, to make a good compofition, are two loads of fludge, one load of foft peat-mould, a load of loam, two loads of rotten dung, and two loads of fharp fand. This is the proportion; the quantity muft be according to the number of rows of afparagus to be fown. If the ground, where the feminary is to be made, is of a light nature, there fhould be two loads of loam and one of fand; but then there muft be more of the natural earth and more rotten dung laid in the bottom.

Afparagus will thrive exceeding well in this compofition, and grow to a great fize, beyond what can be imagined. The trenches fhould be filled with the prepa-red ftuff as foon as they are dug out, which fhould be laid in lightly, and within three inches of the top, that it may fettle equally; it fhould be a foot below the level of the ground between when the feeds are fown. Make a drill in the middle of the trench two inches deep, and at a foot diftance drop four good feeds an inch from one another; prefs them gently into the ground, and cover them with mould two inches deep.

IF all the feeds grow, leave only two of the beft; and if the fpring and fummer be very dry, give them a little water once a week, and keep them clear from weeds. If a thin cover of mofs was laid round them, it would prevent the ground cracking after watering.

VOL. II.   O   IN

In the utumn, when the ſtems are decayed, cut them off two inches from the ground, ſtir the ſurface gently that the crown of the plant may not be hurt, and then lay on an inch and a half of the ſame compoſition that the trenches were filled with, the whole breadth of the two feet, raiſing it a little higher in the middle to carry the wet off the crown of the plants.

It will be neceſſary this winter to cover the whole two-feet trench with light litter from the ſtables, as the roots are near the ſurface ; but this is no more to be practiſed. This finiſhes the firſt year's work.

The beginning of March take off the litter, and dig it into the three-feet left between the rows ; prick up the ſurface in the trench where the aſparagus is ſown, and lay over it an inch more of the ſame compoſt it was ſown in.

There ſhould be as much of the compoſition

pofition made at firft as will fill up the
trenches at leaft two inches higher than
the ground between, which fhould be
turned over twice a year, for it will be
three or four years before it is all ufed.
It is quite unneceffary to repeat the keeping clean from weeds, as that fhould be
always underftood.

In the autumn, when the ftems are decayed, cut them off four inches from the
ground, and lay over the whole trench
four inches of the compofition, raifing it
a little round in the middle.

The crowns of the plants will now be
near eight inches below the furface, fo
there will be no occafion for any long covering ; but as the plants are ftill in danger
from froft, it will be right to cover them
a foot wide with two inches of fharp fand,
which fhould be laid round to carry off
the wet in winter, to prevent its hurting
the buds. This finifhes the fecond year's
work.

In

In the spring the sand should be spread all the breadth of the two-feet trench, and two inches of the compost added to it ; then prick it over with a fork to mix it well with the sand. This work should be done the beginning of March, and let lie until the end of April before it is raked.

In the autumn there should be four inches of the compost laid all over the trench that the asparagus is sown in, which should be two inches higher in the middle than at the sides, so that the crown of the roots will be a foot under the surface, and there will be no occasion for any further cover to keep out the frost. This ends the third year.

The next spring fork the rows, and let them lay a month before they are raked. This spring the heads will be large and fine, and there will be great temptation to cut, but it would be much better to defer it until the spring following.

The next autumn there should be four

inches of the compoſt laid all over in the ſame manner as before, ſo that now the two-feet trench will be an inch or two higher than the ground between.

As the roots will now begin to come to the hard ground between the rows, there muſt be ſome proviſion made to encourage them to ſpread, that they may grow ſtrong and ſupply the buds, which will now be very plentiful and very luxuriant.

ANY time in the winter there ſhould be a trench opened at the end of the three-feet between the rows, the ſame depth that the aſparagus was ſown in. If the ground is light and tolerably good, only a third of the natural earth ſhould be carried off ; but if ſtiff, one half at leaſt.

THERE ſhould be ſix inches of good rotten dung laid in the bottom, then eight inches of the natural mould, and on that a
O 3   layer

layer of dung, and fo on until the ground is four inches higher than the ground the afparagus is in, for it will fink to its own level. If the ground is ftiff there fhould be lefs mould laid above the dung, and an inch or two of fand added; ther it will be neceffary to mix each layer of dung, fand, and mould, by pricking it over twice with a fork.

THE next autumn there fhould be four inches of the compoft laid over the trench, which then will be juft fo much higher than the ground between. If there is any of the compofition left it may be laid all over the ground; but if there is none, an inch of rotten dung will anfwer as well.

IN two years after this it will be necef- fary to fpread an inch of rotten dung all over the ground in autumn, and fork it in, taking hold of the ground three or four inches, and let it lie rough all win- ter; but the middle of the row, where

5   the

the afparagus is fown, fhould always be higheft, and lie round. This fhould be repeated every third year; and by being thus managed the afparagus will laft for twenty years, and be much larger than what is fown or planted in the beds in the common way.

As to the proper age to cut afparagus, there fhould be none cut before the third year; but as the intention of fowing and managing it in this manner is in order that it may be very large, there fhould be none cut before the fourth year; and none even then but a few of the largeft heads, which will give it ftrength to grow to the fize intended.

The many directions here given may feem tedious, and to require fo much labour that it is not worth the trouble: there is not a great deal more than generally is ufed (or fhould be) in preparing of ground for planting in the common way, as trenching and dunging is always

O 4   practifed;

practifed ; for after the trenches are emptied and filled again with the compoſt,
there is much leſs labour in the dreſſings
than is uſed in the common method. But,
as this is a new way, I have not omitted
the minuteſt article in the directions, by
following of which, aſparagus will be
brought to a ſurprizing largeneſs.

I HAVE often heard gentlemen complain that they could never have good aſparagus in the common way, their ſoil not
being fit for its production : this does
not come within my province ; but for
the advantage of ſuch as are ſo ſituated, I
ſhall lay down a method, by following of
which good aſparagus may be obtained,
let the natural ſoil be what it will.

A CLAY bottom is very bad for aſparagus ; therefore where that is the caſe, in
all the ways of ſowing and planting, it
will greatly add to its growth to lay two
inches of coarſe gravel, or ſome ſtones
broken very ſmall, in the bottom of every
trench.

THE bottom fhould be very even, and
have a good fall to carry off the water (if
clay); for if there is not a free paffage it
rots the roots of all plants that are perma-
nent, and it affects none fo much as it
does afparagus; fo that unlefs the ground
is dry at the bottom, and of a good depth
of foil, all endeavours to have it good will
prove in vain.

ALL the beft afparagus grows in light
foils. If the ground is naturally ftiff it
muft be made light; for which purpofe
there fhould be as much fand and coal-
afhes, finely fifted, mixed up as will make
at firft laying twice as many inches as
there is of the natural foil (for fand and
afhes will not raife the ground in proportion to what is laid in, when well mixed
with the mould); the fand fhould be the
fharpeft that can be got; for foft fand,
laid in ftiff clay, only binds it fafter.

MARK out the ground that is intended
to be fown, and open a trench as deep as
the

the ground will allow; lay in the bottom
fix inches of good foft cow-dung that is
not rotten, but has little ftraw amongft
it ; then fix inches of the natural foil,
and on that fix inches of the fand and
afhes, which fhould be well mixed (but
the dung fhould not be difturbed) ; then
lay on three inches of rotten dung, three
of mould, and three of the fand and afhes,
which fhould be all well mixed ; then
three inches of mould, two of dung, and
four of fand and afhes, all well mixed ;
let them remain thus until fowing time.
This fhould be cone the beginning of
winter, and it will be in fine order, and
bring very good afparagus.

IT fhould be fown in drills two inches
deep, and covered with an equal quantity
of rotten dung, fand, and good light
mould, well mixed.

THE beds fhould be covered with light
litter the firft winter ; the fecond, with
two inches of rotten dung, which fhould
be

be forked up, and mixed with the earth on the beds in the fpring; the third winter there fhould be a good covering of dung, fand, and good light earth laid all over, which will be fufficient to keep out the froft.

THERE fhould never be any thing planted in the alleys, neither fhould they be deep; but the beds fhould be laid round, the alleys dug every winter, and fome earth thrown over the beds out of them, the moft of which will be raked off in dreffing.

THEY fhould have at leaft an inch of rotten dung thrown over them every third year, which fhould be forked up in the fpring, taking hold of three inches of the mould on the bed. This fhould be done the beginning of March, and lay rough until the beginning of April before it is raked.

IF the ground intended to be fown be
gravel

gravel or fand, a fine rich loam fhould be
ufed inftead of fand and afhes; but rotten
cow-dung will anfwer better than horfe-
dung; and if all the layers of dung, except the laft, is not fo very rotten it will
be the better.

AFTER all the ground is trenched and
well mixed, there fhould be an inch ot
very rotten dung, two of loam, and one
of good black mould, laid all over, and
pricked over at leaft three times, that they
may be well mixed. This work fhould
be done in the beginning of winter, and
by lying rough it will be in good order
for fowing in the fpring.

IF the natural foil is either a light loam
or a fine black mould, there will be no occafion for any thing except dung; but in
all forts of foils there fhould be a good
layer of dung in the bottom of the trench,
as the lower parts of the beds cannot be
mended after the afparagus is fown; and
when the roots get to the bottom (as
they

they foon will) they will grow ftrong, and greatly increafe the growth of the plants.

IT is a wrong method to plant afparagus, for the roots fhould never be cut, becaufe they are of fuch a nature that they never make fo good roots as when they are fown where they are to remain.

IT is natural for afparagus to rife in the crown as it grows old, for which reafon there fhould be a good covering of earth laid on at different times, that it may be in a condition to refift the froft without a covering of long dung, which is very prejudicial to it, efpecially on ftiff land. Thofe who prepare and manage their beds thus will have good afparagus, let the natural foil be what it will.

ASPARAGUS draws a great deal of nourifhment from the ground every year; it is a very luxuriant plant, and fhould have fome fupply every other year, or it will in a few years be very fmall; but if properly

2   fup-

ſupported it will be very good for many years.

For cing aſparagus, and having it in the winter months, is much in practice : it is ſometimes (when the plants are good) pretty large, and looks tolerably well, but it is impoſſible it can be good ; for the buds are forced out by the heat only, without having any ſupply from the roots, except what juices were in them ; and they ſtrike no new fibres, therefore can draw no nouriſhment from the mould they are planted in ; for when the buds that were formed before planting are all ſhot, there is an end, and the plant decays.

Good aſparagus may be obtained at a trifling expence, from the beginning of March until the beginning of May, of a fine colour and good taſte. This is no part of what I propoſed to treat of ; but, as it is not in common practice, I ſhall give directions how it may be brought to perfection.

BUILD

BUILD a pit thirty or forty feet long, fix feet broad (inward dimenfions) and three feet high. Before the flues are begun the foundations fhould be laid two feet four inches broad; that is, eight inches for the flue, four inches for its infide fupport, and fixteen inches outfide wall, which may be built either of ftone or brick, as moft convenient; but the fupport and cover of the flue muft be brick.

WHEN the wall is three feet high all round, begin the flue, which fhould be two feet deep; and after it is covered fhould have fix inches of ftone or brick built over it before the caping is laid on.

IF the ground this pit ftands on be clay, there fhould be a foot of loofe ftones broken fmall, fpread all over the bottom, and on that a foot of brufh-wood, cut fmall, and a drain through the wall at the bottom, to carry off the wet; but if it is fand, gravel, or rock, two feet of brufh-wood will be fufficient without a drain.

THE fire-place ſhould be ſunk a foot lower than the bottom of the pit, to give room for an aſh-hole below the grate, and ſufficient height to the fire-place, that there may be a riſe of eighteen inches at leaſt before the flue comes to the level, which ſhould go all round, and the chimney ſtand over the fire-place, which ſhould ſtand without the ſquare of the pit.

THE flue at firſt ſhould go along the end, and be carried within four inches of the outſide of the wall, to prevent its being too hot at firſt going off; but ſhould, as ſoon as paſt the end, be brought gradually into the ſquare, to have only the four inches ſupport the remainder of the way round.

WHEN the whole is finiſhed, and the bottom prepared according to the ground it ſtands on, the pit ſhould be filled with the ſame compoſition that was made for the trenches for large aſparagus. This ſhould be done at leaſt two months before

the

the time of fowing; for it will fink greatly, and it fhould be juft a foot below the
caping when the feed is fown.

IT may be fown thicker than that in
the trenches; but five rows will be fufficient in the fix-feet bed, as the rows on
the outfide next the flue muft be at leaft
eight inches from the flue, for the heat
will dry the roots too much if nearer.

THE firft year after fowing it fhould be
covered with litter, in the fame manner
with that fown in the trenches; and managed, as to covering, with the compofition, after the fame manner (which need
not be repeated). When it is finifhed,
which will be the third year, it fhould
lay round the fides, next the flue, an inch
above the caping, and the middle four
inches higher.

IT fhould not be forced before the fourth
year. The fires fhould be made the beginning of February, but very flow at firft,

VOL. II. P and

and increafed gradually. If the feafon is not very fevere there will be afparagus fit to cut the latter end of March.

THERE fhould be a wood frame made like that which fupports the tilt of a waggon laid over, that it may be covered with mats at nights and in cold wet days.

WHEN the fires are made ftrong it fhould have a little water all round the fides next the flues every day, and the whole bed fhould be kept moderately moift.

As the forcing will wafte the afparagus much more than that which grows in common ground, thefe fhould have a little rotten dung laid over them every year, and it would be proper to provide two fuch places in order to have a conftant fupply; for if it is forced every year it will not laft a long time, as the heat will force all the buds to fprout; fo that it fhould have reft every other year to recover ftrength.

THE

THE forcing of afparagus in the winter months, by planting it on hot-beds made of dung, is brought to as great perfection as it can be, unlefs a better method be found, that it may have the advantage of roots and fibres to encourage its growth, as well as heat to force the buds to fhoot, which I hope to accomplifh.

WHERE tan can be conveniently got, it makes a much better bed for forcing winter afparagus than dung; for after the tan is brought to a proper heat it keeps fo a long time, and is not fo apt to burn the roots as dung beds; which, after they have come to their heat, decay much fooner than the bark, and occafions much more trouble in lining, as the heat muft be kept up at that feafon, or all is foon loft.

THE roots of afparagus are foft, fucculent, and fibrous; and when cut, broke, or difturbed, are very fubject to rot: if it does not rot, it pufhes out fmall flender

roots,

roots, which fpread and entangle one amongſt another, ſo that they never make ſuch vigorous ſhoots as when the roots are ſtraight and ſtrong, which they are when ſown and never diſturbed.

THIS is a good reaſon for being at ſo much trouble in preparing the beds before ſowing, alſo for laying ſo great a quantity of dung in the bottom of the trench ; for planted aſparagus will never ſtrike their roots ſo deep as what is ſown.

ASPARAGUS ſeed is very hard and dry, and long in coming up ; yet it is of ſuch a nature, that if ſown very early, and hard froſts or much rain happen after-wards, it is apt to rot ; or if the ſeaſon is mild at firſt, and it comes up ſoon, and a hard froſt ſucceeds, it cuts off the young ſhoots cloſe to the ground, which is a great detriment to the plants, for there are few roots to ſupport them, and a good deal of the ſubſtance of the ſeed is ſpent in produ-cing the firſt ſhoots : thoſe that come
after

after are very weak, and it is a year before they recover their ftrength.

The beft method is to prepare the feed three weeks before the time of fowing, which then may be three weeks longer deferred, by which time the hard frofts will be over, and the feed will be as far advanced as if it had been in the ground in a mild feafon.

To prepare the feed, mix an equal quantity of dry fand and frefh grains from the brewhoufe ; rub them well together, lay it in a heap four or five days, then rub and mix the feed with it, and put it in a little box or garden-pot, covering it over an inch at top with fand, and fet it in a dry airy place ; in fix days examine it, and if it inclines to mouldinefs, rub all over again, and it will require no further trouble, but will be fprouted, and in good order for fowing in three weeks.

P 3   C H A P.

CHAP. XVI.

On Cabbages.

THE cultivation of cabbages for feeding cattle is of great benefit, and its advantages are so great, that it is surprizing that it has not become general long before this, especially as there are several gentlemen that have brought them to great perfection, and found them very profitable.

In several conversations I have had with farmers on this head, most of them made the following objections: That it was too expensive, and that it impoverished the land; that it also required more dung than they could afford. So far they were quite wrong, for it takes no great labour, and

it

it greatly improves the land ; and if they are properly managed, it ferves for a fummer fallow and there will be a very good crop after cabbages without any dung.

If they are well managed and brought to a large fize, there will be more weight on an acre of good cabbages than there poffibly can be on an acre of the very beft turnips. They are eafier to cultivate ; for they are not fubject to the fly, which in dry feafons often deftroys whole fields of turnips : whereas if winter cabbage-plants are planted, they feldom require any water ; and for the fpring plants, if the feafon is ever fo dry, a good watering at planting will fuffice to make them grow.

CABBAGES are of great ufe in deep fnows and hard frofts, when turnips cannot be got ; and I have been informed, that if all the rotten leaves are taken off, and the found cabbages given to milk-cows, that they affect neither butter nor milk. This I cannot affert as a fact that I have feen

P 4   tried,

tried, but I had the information from an ingenious gentleman of veracity.

It is well known by every person that has given turnips to their milk-cows, that they give the butter so strong a taste, that in a few days it is hardly eatable, even by those whose taste is not very nice. In markets it is generally the first question, Is it turnip-butter? and if it is, no gentry will purchase it, and it is always sold cheaper. If feeding milk-cows with clean cabbages will remedy this, it will be an acquisition to the public, and a very great advantage to the farmer.

Cabbages will thrive in all soils except in poor gravel, which is a great advantage to the farmer, as gravelly grounds are fittest for turnips.

In the strongest clays cabbages may be brought to great perfection; and as that is not a soil fit for turnips, there they must be a great improvement; for if most of the

farm

farm is ſtrong clay, and unfit for turnips, in years when hay is ſcarce it will be a great advantage to have a good crop of cabbages.

IF cabbages are intended to be planted on a very ſtiff clay, ſuppoſe bean or wheat ſtubble, the field ſhould be plowed as ſoon as the corn is cleared, and then again juſt before winter, in the ſame manner ; and if the ſpring is dry it will be of ſervice to harrow it well ; but that ſhould be done juſt before it is plowed, for it will become very fine by harrowing after the winter's froſt ; and if much rain ſhould fall before it is plowed, it will run all together and become quite ſtiff ; ſo that it muſt be plowed as ſoon as poſſible after harrowing.

BEFORE the laſt plowing there ſhould be eight or ten loads of good rotten dung ſpread over each acre ; then plowed, well harrowed, and planted immediately.

PLANTS that have been ſown in the autumn,

tumn, and pricked out in beds four inches
afunder, are the fitteft for clay ground, as
they are larger and hardier than thofe that
are fown in the fpring, and they may be
planted three months fooner, which will
be a great advantage, for they grow very
flowly for fome time in fuch ground; but
after they have got good hold they will
grow very faft.

ALL cabbages fhould be pulled and carried to a fold, or fome convenient fhed, to
be eaten, it being very improper to allow
cattle to go amongft them; for if they are
confined to ever fo fmall a fpace they wafte
and deftroy them greatly; and that is
the reafon the farmers fay it impoverifhes
the ground, for the crop is carried off and
fpent in another place; but they fhould
confider what a quantity of dung is got in
the fold, the cabbage ground having no
need of it, which is a great advantage.

A GOOD crop of beans may be had
on clay ground after cabbages without
any dung.

IF it is intended to have beans, the cabbages fhould be all cleared off by Candlemas, and the ground plowed and fown as foon as the weather will permit ; but if after plowing it was well harrowed, and the beans planted with fetting-fticks, there would be a much better crop, and a quarter of the feed would be fufficient.

THIS is practifed in many places in the fouth, where both labour and ground are much dearer than in the north, and they find it turn to good account.

THEY plant them eighteen inches, fometimes two feet, row from row, and three inches in the row. Two women, with a bag in their aprons to hold the beans, plant them very quick. They have two lines beginning at oppofite ends, each planting the whole of their own line, fo that they are ready at both ends to fhift the lines. They make the holes with the fetting-ftick, and drop a bean in each ; and when the field is planted, harrow it all over.

WHEN the beans are two or three inches above ground they ſhould be hoed in the row, to deſtroy the weeds that are in the line of the beans, and then a furrow drawn with a plow between the rows. This furrow ſhould be drawn as ſoon as the beans are hoed, for if they are any taller they will be in danger of being broken. It may lie until the beans are a foot high, and then be drawn up to their ſtems with a hoe.

BY this method of working the ground is kept quite clean, and will be almoſt as good as another ſummer fallow. The goodneſs of the crop will pay all the extraordinary expences triple, beſides the advantage of the ground being kept clean.

IF the ground intended for cabbages be a ſtrong rich loam (which is the very beſt for them) they will grow very large ; but it ſhould be plowed before winter, and lie rough to mellow. If it is to be planted with winter plants, which is alſo the beſt
for

for this foil, it fhould be harrowed, dunged, plowed, harrowed, and planted in fucceffion as foon as can be; but if it is to
be planted with fpring plants, it fhould be
harrowed as foon as tolerably dry in the
fpring, and plowed directly; for if it is
not plowed before planting time it will be
full of weeds.

Just before the feafon for planting,
which fhould be as foon as the plants can
be got to a good fize, it fhould be dunged,
plowed, and harrowed, and then planted.

Little dung will be fufficient for fuch
ground, and a good crop of barley will
grow after the cabbages without dung.

If the ground to be planted with cabbages be of a fandy nature, or a light
loam of a good depth, they will thrive
extremely well. It would be better to
plow it before winter; but if plowed early
in fpring it will anfwer.

Ground of this nature anfwers the beft

of any for fpring plants, it being naturally foft, and they will take root much
fooner than in any other kind of foil, and
it fhould be dunged and plowed juft be=
fore it is intended to be planted.

AFTER it is plowed it fhould be well
harrowed, and then a fhallow furrow drawn
by the plow at four feet diftance from one
another, and the cabbages planted in the
bottom of it at two feet and a half diftance.

WHEN the cabbages are grown to have
three or four inches of ftem, the weeds
fhould be hoed between the plants in the
rows, and the loofe earth that was thrown
up in making the furrow, fhould be drawn
clofe up to the ftems.

ALL the cabbages that are planted in the
different grounds that were winter plants,
about the end of June fhould have two
furrows drawn by the plow between every
row, throwing the mould up to the cabbages; after which it fhould be drawn

3   quite

quite clofe to the ftem by a hoe. It will
be the middle or latter end of July before
it will be neceffary to draw the furrows
and earth up the fpring plants.

WHEN cabbages are planted in a fine
light fandy foil that is of a good depth,
if the ground is kept clean, and the
cabbages cleared off by Candlemas, the
ground may be put in good order to produce a fine crop of carrots.

WHEN carrots are intended after cabbages, the cabbages fhould be pulled up
by the roots, for their roots would be very
troublefome in the carrot ground. The
ground fhould be plowed as deep as poffible and lay rough.

THE feed for winter plants fhould be
fown at two different times, for there is
often three weeks difference, or more, in
the feafon. The firft fowing fhould be
about the latter end of July, and the
fecond about the 18th of Auguft.

IF

If the autumn is very fine, which often happens, those sown first will be too large; and if they stand through the winter many of them will run to seed in the spring; but if the autumn prove cold and frosty, as it sometimes does, those last sown will be too small and weak, and unable to resist the cold winds and hard frosts.

When the plants have got four leaves they should be pricked out into beds in an open spot in the garden or in the fields (where hares and rabbits cannot come at them) at four inches distance every way. If they are planted near walls or hedges it draws them, and they have long stems, which makes them weak, and is a great detriment; for it is the strong stiff plants which make the large cabbages.

All those that are the least acquainted with gardening know, that if the season is dry, all young plants, when first planted out, should have water, and it would be very superfluous to direct it; but this is

5   designed

defigned for the farmer (not the gardener) who is fuppofed not yet to have come to a perfect knowledge of managing plants of this kind.

IF the feafon is very dry when the young plants are planted out, they fhould have a good watering in the evening or morning after planting; but if ever fo dry fhould have no more. It is much better for them to be ftiff and fhort, both for ftanding the winter and planting out in the fpring, than thofe that are flufh, tall, and tender, which watering or a wet feafon makes them.

THE ground which the young plants are pricked out in to ftand the winter, fhould be a good natural foil, not made rich with dung; for then the plants grow too faft, and are fo thick that they are drawn; and if it comes a hard winter they are in danger of being loft.

IF there are no winter plants, and the whole is to be planted with spring plants, they should be sown as soon as possible, in a spot of good ground that lies pretty warm, to bring them forward.

THEY should not be sown too thick, and it would be worth the labour, after they are come up, to thin them by hand to four inches distance, as the trouble would be all saved in the planting, besides the advantage of the plants being much larger; for when they are very thick they are long stemmed, crooked, and very troublesome to plant, which they are not when thin.

THE season for planting plants that have been kept over the winter, is any time from the beginning of March to the end of April; but the sooner they are planted the crop will be the better; for in that early season they have a good chance to get rain after planting; and as the ground is for the most part moist, they will re-

quire

quire no watering when planted, which is faving more labour and expence than the tranfplanting of the plants cofts in autumn, befides the advantage of the plants having almoft three months longer to grow.

IF fharp frofts fhould enfue after the cabbages, are planted in the fields, it will not hurt them, for the moving makes them more hardy than if they had not been tranfplanted.

IT will be the beginning or middle of June before the plants that are fown in fpring are ftrong enough to plant out. It is wrong to plant them fmall in the field, for if the ground is not fine it will be very difficult to make them faft, and at that feafon the weather is generally hot. If they are not good large plants they will ftand a bad chance, although they are well watered when planted.

If the weather is rainy and dull, it will be much better for them ; and if there is an appearance of rain it will be right to wait a few days; for it is much better planting if the ground is wet at top, as it prevents the mould running into the hole, which is very troublesome in planting.

If a good deal of rain should fall just at the time of planting (but it very seldom happens that a sufficiency falls at that season to moisten the ground deep enough for the roots of the plants) it would even be a great service to them to be watered when the weather is dull. One quart will be of more advantage than a gallon when it is hot and dry; and if in a few days it should be very hot, and continue so for a long time, they will require no more watering.

If the ground has been well prepared, and is very tender and fine, it will have a good deal of moisture in it; and if some rain has fallen they may do without water; but

but if the ground is rough and lumpy, water is abfolutely neceffary, although there has been a good deal of rain.

If cabbages are managed on the different foils, as here directed, they will be a profitable crop, and few will be without them after they have experienced the great advantage they are of in winter-feeding.

No cattle are fond of eating vegetables when they are frozen; therefore if the cabbages are carried home and laid in a cow-houfe fingly, where cattle are kept, the heat of the place will foon thaw them; but if they lie in a heap they will continue frozen, although in a very warm place.

It will require no great room to thaw them; for after a few are thawed, and the cattle fed, fome may be laid all along behind them, to be ready againft thefe are eaten. This will be very little trouble, the cattle will be much better fed, and there will be no wafte.

Q 3 CAB-

CABBAGES may be taken up and laid close together in the earth and covered with peafe or bean ftraw to keep the froft from them ; that light covering will admit the air and prevent their rotting.

THE feeding in winter with vegetables that are not frozen muft be a great advantage ; they are more nourifhing and make the cattle much fatter and fooner.

CHAP.

C H A P.   XVII.

On Carrots.

CARROTS are another very beneficial vegetable for feeding, and many other profitable ufes. The cultivation of them in all the different foils they will thrive in will be very advantageous to the farmer. I fhall endeavour to give fuch plain and eafy directions for the propagation of them, as I hope will be fuccefsful, allowing for the common accidents of feafons.

ALL the farmers may keep their horfes from the beginning of September to near May-day with carrots and hay, without one grain of corn, which muft occafion a great faving of oats.

THE

THE following is an abſolute fact: That three acres of carrots, on a ſoil that is not the very beſt for the growth of them, ſupported twenty-four working horſes, thirty grown ſwine, and ſeveral young horſes and cattle, from the end of October to the beginning of April. The horſes were worked every day, looked well, and were fat.

BESIDES there were quantities frequently given to the milk-cows during that time, which had no bad effect on the milk; and they gave a greater quantity at a meal than if they had not had carrots, and the butter was ſweeter than of thoſe which were fed with the beſt hay only.

THE labour in keeping carrots clean is not much different, nor more expenſive, than that of turnips, where turnips are hoed and kept clean as they ſhould be.

TURNIPS will be ſomething like a crop, although they are not hoed, and are full of weeds, and there are many places

5 where

where fuch flovenly work is feen ; but the occupiers of the grounds fuffer for their negligence, both in the lofs of turnips and the dirtinefs of their grounds, as the feeds blow all over the farm ; but unlefs carrots are kept clean and hoed to a proper diftance, they will be good for nothing.

I AM of opinion that there are few farmers, after they have found the great advantage carrots are of in feeding, but will endeavour to have fome acres of them, which requires no great art, if the whole farm is not a hard clay or a fhallow gravel; in that cafe it is in vain to attempt their propagation.

CARROTS will thrive on fandy grounds of all kinds ; light loams, ftrong loams, light black moulds, and all kinds of land that is of a loofe, open nature,

THE ground which carrots are fown on fhould never be dunged that year ; but
they

they fhould follow a crop that has been dunged the preceding year; for the ground muft be in good heart, or there will be little profit.

THERE is no crop that prepares the ground fo well for carrots as cabbages; when carrots are intended after cabbages, before the cabbages are planted the ground fhould be well dunged previous to the laft plowing.

THE cabbages fhould be all got off by Candlemas, and as foon as the ground is tolerably dry it fhould be plowed as deep as the plow can go, and lie rough until near the time of fowing: it then fhould be well harrowed with a large harrow, which will make it fine at top. It muft be plowed again directly, for if rain fhould come before it is plowed, all the fine mould at top will run into a hard cruft, and all the advantage of the early plowing will be loft.

THERE

THERE are some kinds of soils (which have been mentioned before) that cabbages will thrive on, that are not fit to be succeeded by carrots; but as carrots are so advantageous a crop, if there is any ground in the farm that is fit for them, it should be so contrived as to be planted with cabbages, that the carrots may be sown after them.

IT is a good method, where there are large quantities of potatoes propagated, to plant first potatoes, which are always dunged with long dung.

To prepare the ground properly, it should be plowed as soon as the potatoes are taken up, to mix and rot the long dung that they were planted with. It should lay rough all winter; early in the spring harrowed, then dunged according as the ground is in heart; then plowed, harrowed, and planted with cabbages, to be managed as before directed, and sown with carrots next spring.

ALL

ALL thofe crops will thrive on the fame fort of foil, and produce good crops of each kind.

IF the crops are fown in this fucceffion, the potatoes fhould be kept very clean, and no weeds allowed to feed on the ground; for if they do they will be very troublefome in the carrot ground.

THE cabbages the next year muft alfo be hoed and kept clean, and well plowed between the rows, which is almoft as good as a fummer fallow; by this ma-nagement the carrot ground will be in fuch good order next fpring that the hoeing will be eafy, and the crop good.

THE beft foil for carrots is the black deep fandy grounds that are in low bot-toms by river fides; there they grow large and fine, and prove a very profitable crop. If fuch ground can be had, the labour will be trifling. Suppofing it had produced cabbages the year before, it fhould be
plowed

plowed when the cabbages are got off,
and lie rough until the time of fowing the
carrots, and fhould then be harrowed with
a large harrow, plowed immediately, and
harrowed, the feed fown, and then flightly
harrowed with a bufh-harrow.

IF the ground intended to be fown be a
deep but hungry fand, it fhould be managed in the fame manner, but fown as
early as poffible; for if the plants do not
get good hold of the ground before the
dry weather comes on in the fpring, they
will come to nothing; but if fown early,
there will be a fine crop.

SANDY loam is a very good foil; fuch
ground is always of a good depth, and a
fine crop may be expected. It fhould be
treated much in the fame manner as the
former, only it will require a little more
harrowing before the feed is fown.

IF ftiff loam, carrots will thrive very
well; but it will require a good deal more
working

working to reduce it; for rough ground is very unfit for carrots.

IF ſtiff loam has not produced cabbages the year before, it would be much better to be fallow for a ſummer; and if it is not in good heart, to have five or ſix loads of very rotten dung per acre ſpread over it in ſummer, and plowed in directly. It ſhould be kept very clean.

THIS method will be much condemned by the generality of the farmers as too expenſive, on account of the two winters and one ſummer's working and dunging alſo for one crop; but I am ſure it will pay very well, for it will produce a very good crop if the ſeaſon proves temperate.

IF there is a more proper ſpot I would not recommend ſuch ſoil, only in caſe of neceſſity, that it is the beſt in the farm, that thoſe who have not a fitter ſoil may not deſpair of having good carrots.

BLACK

BLACK mould, and all loose earths, will answer very well for carrots ; but they should be sown early, for the same reasons that were given for sowing sandy ground.

IF the best ground in a whole farm is all stiff and next to a clay, it is very unfit for carrots; but at a small expence it may be made to grow them tolerably well; and the ground will also be greatly improved for all kinds of grain, and last many years.

THE following composition, prepared and laid on, and plowed and harrowed two or three times afterward, will mix with the natural soil, and bring it into good order for sowing. If the season is not very wet there will be a good crop.

TWELVE loads of sharp sand, four load of light loose earth, and one load of rotten dung, all well mixed, and turned at least twice before it is laid on. Twelve loads to an acre will do pretty well ; but

to

to make the ground very good it would take fixteen.

ALL the expence of this muſt not be reckoned to the account of the carrots, for next year a good crop of turnips may be had, if the ground is plowed and laid rough all winter, and properly worked in the ſpring, and ſown at the uſual time.

THE ground that is intended for carrots ſhould always be plowed in the autumn, and lie rough all winter; but if they are to ſucceed cabbages, theſe ſhould be all got off by Candlemas, and the firſt time the ground is tolerably dry it ſhould be plowed, that it may get ſome froſt to mellow it.

IF the ground intended for carrots is of a ſtiff nature, it ſhould be well harrowed with a large harrow every time juſt before the plowing; but it ſhould be dry when both harrowed and plowed ; for wet plowing is very detrimental to all
ground,

ground, but much more fo to ground where carrots are to be fown.

The reafon for fo much harrowing is, that the ground may be all very fine the depth of the plow; for if the furface be only fine and rough below, the clods will ftop the fmall roots of the carrots in their going down, and make them forked and good for nothing.

If the ground was quite hard it would be much better for them than rough; for, when rough and loofe, the fmall end, as foon as it meets with the leaft obftruction, (if the mould is loofe) the root bends, and fo grows into many forks; whereas if the ground is all equally hard and free of ftones, they will ftrike a great depth into the hard ground.

The ground fhould never lie any time after the laft plowing before it is fown; for if dry weather fhould fet in, the feed will lie a long time before it comes up and a great deal of it will not appear until there is rain.

THE weeds at the same time will grow, and there is no hoeing until the carrots are all up: in that case the largest of the weeds must be pulled by the hand; for if they are permitted to grow until the carrots are visible, they will be too strong to cut with the hoe; and if then pulled they will draw many of the carrots out of the ground, for there are many of the annual weeds that spread their roots very much when they are permitted to stand to grow large.

WHEN the season will permit, the earlier the carrots are sown the better: if the ground is tolerably dry, they may be sown the beginning of March, for after they are come up they will stand a hard frost without sustaining any injury.

THEY should be sown as soon as possible after the ground is plowed for the last time.

time. The method to proceed is this:
plow and harrow; then sow and harrow
again.

If the soil is stiff (which it will not be
if it has been managed as directed) a heavy
harrow should be drawn over it three or
four times before it is sown, and then harrowed twice over with a light harrow after
it is sown. A pound and a half of seed
is sufficient for an acre; but those who
have not been accustomed to sow it had
better use two pounds. They are easily
cut out in hoeing, if too thick.

The seed being very light, the husks
that adhere to it make it very apt to stick
close, and render it hard to separate; therefore the best method is to mix it with
twice as much sand as seed, of a different
colour from the land that it is sown in; if
the land is of a red colour use white sand,
and red sand, if the ground is of a whiteish
colour; rub it between the hands until the
husks are mostly broken off; then the

feeds

feeds will fly more regularly, and the fand will fhew where the ground is fown. It is more difficult to fow than turnips, but a little practice will make it eafy.

The beft method of fowing carrots is to prepare the feed before fowing; and then, if fown as directed, it will be all up as foon as the annual weeds, and may be hoed before the weeds come to any height, which is of great advantage, and faves much labour.

To prepare the feeds: Take three parts dry fand and one of frefh grains from the brewhoufe, mix both together, rub it all over between the hands feveral times, and then lay it in a box.

This muft be repeated daily for four or five days; then it fhould lie four days, and be rubbed again; in four days more it fhould be examined, and if there be no figns of mouldinefs it fhould be covered an inch thick with dry fand, and remain

for

for fifteen days, by which time it will begin to fprout, and be in proper order for fowing.

IF there be any figns of mouldinefs when the feed has lain eight or nine days, (for it will be right to examine it about that time) it muft be all turned out of the box and rubbed over; and if it appears to be wet, it fhould be fpread a few hours in an open airy place, but not in the fun, and then put up again, and afterwards there will be no danger of its moulding.

As foo as the carrots are fairly up they fhould be hoed with a three-inch hoe. If the weather is favourable they will grow very faft, and in fix or eight weeks will want hoeing again, which may be performed with a fix-inch hoe, and left at that diftance.

Two hoeings will be fufficient for the prepared feeds, if the ground be in tolera-

R 3

ble

ble good order ; for before any weeds can
grow after the fecond hoeing, the plants
will cover all the ground and prevent
them.

WHEN the carrots are taken up they
may be laid in any dry place ; but where
there is a quantity it would be beft to build
a fhed or hovel on purpofe for them. It
may be a double roof, or a fhed againft a
wall ; if againft a wall, it muft be well
fecured at top between the wall and the
fhed, for the wet is apt to come in there,
and run down the rafters, and fpoil the
whole.

A DETACHED hovel is the beft with a
double roof, the fide-walls of which fhould
be about fix feet high, and the breadth of
the houfe ten or twelve feet ; for if it was
much broader there would be too great a
body of carrots, and they would be apt
to heat. The fide-walls would be better
if made of whins very clofe, as they admit

mit a free air and keep out the froft; for a wall would make it very damp.

THE roof fhould be thatched, as it will keep out the froft much better than any other kind of covering. The floor fhould be a foot higher than the ground round, in order to keep the carrots dry; and if the ground is wettifh, it fhould be raifed with dry rubbifh, that there may be no damp at bottom.

THE fhed, or hovel, fhould be got finifhed in the beginning of the fummer, that all may be perfectly dry; and as foon as poffible a good quantity of fand laid in, in dry weather, the fharper the better.

THERE can be no particular time fixed for taking up the carrots; but as foon as the tops begin to grow yellow it fhould be done.

THE weather fhould be dry and fine; for if they are taken up wet it will be

R 4

very

very bad for them.   They fhould never be taken up when it is frofty.

A NUMBER of men fhould be employed to take them up with dung-forks, and women fufficient to cut off their tops at an inch from the carrot, and put them into carts to be conveyed home.

IF the ground is light and fandy, and the weather dry, they will be very clean; but if the ground is of the loamy kind, the mould adheres to them, which fhould be rubbed off with the hand, and care fhould be taken not to break the rind. Thofe that are cut or broke, fhould be thrown by for prefent ufe.

WHEN they are brought home there fhould be an inch of the dry fand fpread over about fix feet in length of the floor, and fome very dry wheat-ftraw laid up at the end and both fides of the houfe; then the carrots fhould be brought in and laid even on the floor a foot thick, and a pile

of

of them made ftrait up at fix feet diftance from the end of the houfe.

A LITTLE fand fhould be fprinkled in amongft them, and fome at both the fides; then another foot of carrots and fand, as before, until they are within a foot of the top of the houfe. The laft layer fhould be covered two or three inches with fand.

WHEN that layer is finifhed another fhould be begun in the fame manner, and carried up after the fame method, until the houfe is filled, or the carrots houfed, taking care to lay fand on the floor at the beginning of every layer, and to fill up the fides of every layer with fand, to pro-tect the carrots from froft and air. If the fide-walls of the hovel, or fhed, are made of whins, the air will pafs all round the fand that is laid on the outfides, and keep the whole dry.

IT will be neceffary to have two doors in the fhed, or houfe, if it is of any length,
for

for the conveniency of housing the carrots;
for it would be very troublesome to go the
whole length of the house with every
basket of them.

CARROTS laid up in this manner will
keep very well till May, and be as firm
and good as when first taken out of the
ground.

THE horses that have not been accustomed to feed on them will be very nice
at first, and will not eat them; therefore
they should be cut small, and mixed with
their corn, and they will soon grow so fond
of them as to leave corn and feed on the
carrots.

THERE will be no occasion to cut them
after the horses have taken to eating them.
Cows and young cattle will feed on them
very freely at first. The swine want no
invitation, and for them they want no
washing.

3   UN-

UNFORESEEN accidents fometimes happen, fo that there may be poor crops of carrots; this is common to every thing that is fown; and although all the directions are followed, and the foil proper and in good order, the feafon may be fo bad and contrary to them that there may be few; but this will feldom happen.

ACCIDENTS of this kind often happen in all the common courfes of gardening and hufbandry that have been in practice for many years, and allowed to be a good and fubftantial reafon; but in things that are new, if the feafon is ever fo bad and contrary, the fault is laid to the new projects as being of no utility, only fpeculative, and without any real exiftence but in the brain of the projector.

LET no fuch prejudices as thefe affect thofe that intend to make trial of growing carrots; for although they fhould be unfuccefsful in the firft, and even in the fecond attempt, let them perfevere with refolution,

folution, and they will find their labour well rewarded, for they are in general a more certain crop than any thing that is fown for feeding, and of greater profit than any one can conceive that has not tried them.

THE truth of this I can affert by what I have feen, and by what has been done at Sir Thomas Gafcoigne's, at Parlington, where they are the moft advantageous and profitable vegetable for feeding cattle that is fown or planted in England.

GREAT advantage may be had from the cultivating of carrots in large quantities, as they feed working horfes, cattle for fattening, young cattle, and fwine, without corn, which muft greatly leffen the confumption of oats.

THE kinds of carrot-feeds proper to fow in fields are the large red-horn carrot, and the long large orange.

IF

IF the ground is shallow the horn-carrot is the only seed fit to sow; they will go down as far as the soil will permit, and grow to a great thickness, so that many of them will weigh four or five pounds, and even in such ground will be a very profitable crop.

IF the ground is sandy and deep, the large orange carrot is the best kind. Although it is a hungry, poor, sandy soil, if it is sown early it will produce a good crop, if it has been managed according to the directions here given.

IF the ground is a good, deep, sandy loam, such as the soil is in general about Pontefract, in Yorkshire, they will grow to an immense size, and produce a crop advantageous beyond imagination.

IN such ground, if hoed to a proper distance, which should not be less than ten inches, and kept clear of weeds, they will grow to be ten and twelve pounds weight.

CHAP.

CHAP. XVIII.

*The Cultivation of Turnips, and the Method
to keep them from Frost in Winter.*

THE cultivation of turnips is fo well
known all over England, that it may
feem fuperfluous to give any further inftructions with regard to their culture:
yet, notwithftanding they are already
brought to great perfection, there are ftill
fome particulars that are not generally
known, which, if put in practice, would
render that ufeful vegetable of double the
value it is at prefent eftimated, for the
feeding of cattle.

I SHALL endeavour to give fome ufeful
directions refpecting the fowing in drill
and broad caft, and the proper foils for
this kind of hufbandry. Dry, light,
gravelly

gravelly foil is the beft kind for turnips, in which they fhould be fown broad caft, and never in drills; for the ground being naturally dry, if thrown up in ridges it becomes fo light, that if the feafon fhould prove very dry there will be little profpect of a fuccefsful crop. The more the ground is worked, and the finer it is made, the moifter it will be, although plowed when very dry, and a great drought follows.

Dung laid on light ground for turnips fhould be fo rotten, that it will fpread like afhes. The beft method to fow fuch ground is to harrow it firft, then plow it, and fpread the dung as you would lime or afhes: this done, give the whole a triple harrow, which muft be very heavy, to incorporate the dung properly with the mould. Sow the feed, and give the ground a fingle turn with a light harrow, drawn with fmall thorns, to prevent the harrow making drills, leaving a fmooth furface. If the weather is hot and dry, and

and the ground light, it would be of confiderable advantage to the turnips to roll
the ground a day or two after the fowing.
If the ground is very ftrong, drilling will
be a more certain method for a good
crop than any other I know in practice.
When that is the cafe, the land muft be
plowed and harrowed frequently; it fhould
be harrowed after every plowing, but not
till juft before it is to be plowed again;
for if it was to be harrowed foon after the
plowing, and much rain was to fall, it
would be ftiffer the next plowing than if
it had not been harrowed at all.

Is the ground is very rough after the firft
plowing, which fhould always be performed when the furface is dry, and a
week or two of fine dry weather fhould
follow, and then a good deal of rain, the
day after, if it was well harrowed with a
heavy harrow, it would break the earth very
fine; but it muft be plowed immediately,
for if it was to rain much, it would become a folid mafs, and it would be impoffible
to get it into order in proper time.

It is proper to fow after a thorough harrowing. This management will alfo anfwer for broad caft. If the ground is very rough, the day after plowing, it muft be harrowed with a heavy harrow, and then rolled with a fpike roller, which will break many of the clots.

If the ground is not fubject to wet ftanding on it, in the winter the drills fhould not be raifed more than three inches above the level; for in winter when the fheep are wet, many of them would be loft when the drills are high. If the land is liable to have the wet continue on it after rain, the drills fhould be raifed in proportion, and then the turnips fhould be pulled, and the fheep fed in a dry field. It muft be obferved, that fat fheep fhould not be fed on fuch ground.

The diftance for horfe-hoeing and plowing between the furrows or ridges fhould be two feet; but if the ground is to be cleaned with the hand hoe, they

Vol. II.   S   may

may be from fixteen to twenty inches. The horfe-hufbandry is preferable where the land is ftiff, as the plowing and turning the land, that is to be fowed with turnips the next year, makes it fitter for fowing ; befides it is full as good as a fummer's fallow.

WHERE the horfe-hufbandry is ufed, and the diftance fo great as two feet, it is alfo abfolutely neceffary to have them two years in the fame field, that is, where the opening between was the firft year, fhould be the place for the drill the following year ; by this means the field is equally dunged, and all of it has two years fallow.

WHERE this method is practifed on ftiff land, and the land kept clean, it is of greater advantage than can be conceived, and the future crops will amply repay the expence and trouble.

THE dung for fowing turnips in the drill way fhould be very rotten, and turned
until

until it is fo fine that it may be fown by hand into the drill; and if fome good fand was mixed with it, and well incorporated, it would be of great fervice to the crops of turnips, and a great benefit to the land afterwards.

WHEN the land is very rough, rolling it with a fpike-roller would break many of the clots, which fhould be performed the day after plowing; let it be harrowed well with a heavy harrow, and then rolled; plow it again directly after rolling.

IT is a common error to make fpike-rollers too heavy. If the ground is very rough (which is generally the cafe when they are ufed) it is very difficult to draw them; befides they prefs the hard lumps that will not readily break into the ground like fo many ftones, which will be found very troublefome the next plowing.

THE hoeing and keeping clean the ground, where turnips grow, fhould never be neglected. Some content themfelves

with

with pulling out all the great weeds that are above the turnips, and would prevent their growth; but pay no regard to the fmall ones, even with the turnips, and fo full of feed that they ftock the ground with annual weeds for feveral years.

THERE has been a great deal written about the cultivation of turnips, but little faid concerning their prefervation, which is an article that would be of immenfe profit if properly attended to, that is, to fave them from the froft.

THE beginning of long and fevere frofts in general are moderate at firft; fo that the cattle feed on them with eafe for fome days before the turnips are frozen through; in that time the tops of many of them are bit, the froft continues, and often a fall of fnow comes, which lays on the ground perhaps fome weeks. As foon as there is a thaw all thofe that were bit rot directly, and many of the largeft turnips alfo (although not bit) which are

in

in general fpungy, and when once frozen through they foon rot.

WHEN the turnips are come to their
full growth, on a fine dry day pull up all
the largeft, and cut off the tops an inch
from the turnips for the fheep to eat, for
if they are cut clofe they are fubject to rot.

IF this method is put in practice there
will be few of the turnips loft by rotting,
and fave hay in the fevereft weather,
which will be a great advantage in years
when that article is fcarce ; the fmall ones
will refift the feverity of the winter, and
be good feeding late in the fpring : the
turnips will be of great value, as they
will afford fufficient food through the winter; but few muft be given at a time, that
they may be all eat before they are frozen.

FROM November 1776 to March 1777,
although the winter was not fo fevere as
many I have known, yet the frequent
changes from froft to fnow and rain
6   rotted

rotted one half of moſt fields of turnips that were large. If the ground is dry, a large pit dug and covered with ſtraw to prevent the wet and froſt will keep them : a thatched hovel will anſwer, or they may be made into a long or round ſtack, but that is not ſo convenient for taking them out, being ſo very liable to tumble down.

THE ſheep ſhould be fed in the ſevere winter weather on dry graſs fields in the fold, and if the ground is ſtiff where the turnips grow, it would be better not to feed on it in the ſpring.

IT is a great profit in hard froſts and deep ſnows to have plenty of turnips to feed with ; and what makes it ſtill more profitable is, that moſt of thoſe turnips that are then feeding and fattening the cattle would have been rotting in the field as ſoon as the thaw came, if they had not been ſecured from the froſt. According to the preſent method of feeding ſheep on turnips, when there comes a deep ſnow they

they muft have hay, and the turnips they
fhould be eating are rotting in the field,
and are of little fervice to the land they
rot on. If the cattle are fed on the largeft
of the turnips in winter, the fmall ones
that are growing will not run to feed fo
foon in the fpring as the large ones would
have done, fo that there will be food for
the cattle much longer, and between the
turnips, and the grafs, there will not be a
fcarcity of food for fheep; thofe that were
laft pulled would laft until the grafs was
grown, which would be a great advantage
to the farmers.

THE advantages in managing the turnip crop, as here directed, are many, and
eafily put in practice; the expence of
hoeing is trifling, and very little more
trouble than feeding with hay, and the
profit to the farmers will be very great, if
the whole is put in practice.

F I N I S.

Printed in the United States
By Bookmasters